Alfred P. (Alfred Payson) Gage

Solutions of Problems in Gage's Elements of Physics

Alfred P. (Alfred Payson) Gage

Solutions of Problems in Gage's Elements of Physics

ISBN/EAN: 9783743687080

Printed in Europe, USA, Canada, Australia, Japan

Cover: Foto ©berggeist007 / pixelio.de

More available books at **www.hansebooks.com**

SOLUTIONS OF PROBLEMS

IN

GAGE'S ELEMENTS OF PHYSICS.

ALSO

A GENERAL REVIEW, TEST QUESTIONS, AND HINTS TO TEACHERS.

BEING

Parts III., IV., and V. of his "Physical Technics,"
(Published by the Author).

BOSTON:
PUBLISHED BY GINN, HEATH, & CO.
1885.

Entered according to Act of Congress, in the year 1884, by
ALFRED P. GAGE,
in the Office of the Librarian of Congress, at Washington.

J. S. CUSHING & CO., PRINTERS, 115 HIGH STREET, BOSTON.

Part III.

GENERAL REVIEW OF PHYSICS, WITH HINTS TO TEACHERS.

CHAPTER I.

PROPERTIES OF MATTER.

§ 1, p. 1. The pupil may be informed that, though the definition here given of an *experiment* harmonizes with its etymological signification and with the general spirit of this book, yet it is frequently used as a synonym for *illustration* or *proof*.

P. 3, Exp. 5. The apparatus furnished for this experiment is so sensitive that sufficient air can be drawn from the globe by suction with the mouth to cause the beam to tip perceptibly. A change in weight will be still more perceptible if, after exhausting the air by suction, a person were to condense air in the globe by blowing into it, and then closing the stop-cock.

§ 4, p. 4. "The molecules of the same substance are all exactly alike, but different from those of other substances. Molecules are unalterable by any of the processes which go on in the present state of things, and every individual of each species is of exactly the same magnitude as though they had all been cast in the same mould, like bullets, and not merely selected and grouped according to their size, like small shot.

They are, as we believe, the only material things which still remain in the precise condition in which they first began to exist." — MAXWELL.

§ 5, p. 6. The pupil may be aided in applying this theory to the condition of things supposed to exist, for instance, in the test-tube of § 4 filled with water, by imagining that he is looking at a dense flock of birds in the air. The flock remains at rest, but the individuals which compose the flock are in rapid motion, frequently hitting and bounding away from one another. At the same time he sees the whole atmosphere filled with birds of a different hue. These birds, like the former, are in constant motion. Some are constantly entering the space occupied by the flock, and some are leaving it, while all are jostling against one another. The latter birds are nowhere nearly so thick (i.e., there are not so many in a given unit of space) as those which compose the flock, but they are thicker in the space occupied by the flock than in the space outside the flock, partly because their escape from this space is impeded by the presence of the birds of the flock, and partly because of some mutual attraction or affiliation.

The birds constituting the flock may represent the molecules composing the body of water in the test-tube; while the birds of different hue may represent the air molecules. Beware of the common error of supposing that the latter are *smaller* than the former, as in the oft-given but fallacious illustration of "apples, marbles, and bird-shot." Apprise the pupil of the fact that air in water is in a comparatively condensed state; in other words, that a tumbler filled with water contains more air than it would contain if no water were in the tumbler.

§ 7, p. 7. LAW OF AVVOGADRO. — *All gases (at the same temperature and pressure) consist, within equal volumes, of equal numbers of molecules.*

Exp. 2, p. 10. This may well be made a home-experiment, and performed as follows: On two piles of books about 10^{cm} apart support a tin basin containing about a liter of ice-water and a few lumps of ice. Between the piles and under the basin place a lighted candle, so that the tip of the flame will just touch the basin. The books, besides serving as a support, will protect the flame from currents of air. In 10 or 15 minutes the bottom of the basin, except a small area immediately over the

CHAPTER I. — PROPERTIES OF MATTER. 133

flame, will be covered with large drops of water, and a stalactite of carbon will depend from the basin immediately above the flame. After the pupil has performed this experiment he should be plied with such questions as follows: —

Whence came the water and the carbon? What purpose does the ice-water serve? Whence comes the water frequently found on sweating ice-pitchers? How do you know that the water found on the bottom of the basin did not come from the same source? Would the *heat* of the flame tend to increase or retard this condensation? Does carbon always rise from a burning candle flame? Why does it collect in such abundance in this case? What are the ashes of the candle?

§ 11, p. 12, Exp. 1. Inasmuch as the success of this experiment depends frequently upon the condition of the apparatus and the state of the weather, the following substitute will be found very convenient: Take a cork about 1^{cm} in diameter, and cut it transversely into slices about 1^{mm} thick. Then take four cambric needles, break them in the middle, rub each piece two or three times in the same direction across the same pole of a magnet, and thrust each piece perpendicularly through a cork slice, leaving one extremity level with the surface of the cork. Be sure that the ends which are inserted in the corks are all the same poles, *i.e.*, all S. or all N. poles. Fill a bowl or goblet with water to the brim, and float the slices on the surface of the water, with the projecting part of the needle immersed in the water. Holding a bar-magnet vertically, approach the center of the surface of the water with one of its poles, and the slices of cork will sail along the surface, either collecting under this pole of the magnet or receding from it. Reverse the pole of the magnet, and the phenomenon will be reversed.

§ 17, p. 18. The native of Borabora who called hail " white stones " spoke the simple truth. Ice is *stone, a mineral*, in every sense which these terms imply. In the year 1739, at the wedding of Prince Gallitzin, the Russians built for him a house of large blocks of stone. All the furniture of the house, even to the nuptial bed, was made of the same stone; and the cannon and mortars which were fired in honor of the day were constructed of the same. The mineralogical name of this stone is — ice.

§ 17, p. 20. "According to Pictet, oxygen is liquefied at 320 atmospheres pressure and − 140° C.; and then, upon allowing a jet of this liquid to escape into the air, the escaping jet of liquid oxygen becomes extremely cold and is partly solidified. Hydrogen treated in a similar manner, under a pressure of 650 atmospheres, appeared as a steel-blue stream of liquid, the light reflected from which, being partly polarized, revealed the presence of solid particles in the liquid, while the tube of exit became blocked with solid hydrogen." — DANIELL.

Full accounts of the liquefaction and solidification of the "permanent gases" can be found in the Popular Science Monthly, Scientific American, and Nature, of the year 1878.

§ 21, p. 22. It may be demonstrated geometrically that a particle placed anywhere, as A, A', A'', etc., Fig. 18, within a homogeneous spherical shell will be in equilibrium. Consequently, if the earth were such a shell, a body placed anywhere within it would remain at rest. Hence, it follows that if the earth were a homogeneous solid sphere, the weight of any body within it, as B, Fig. 18, would vary as its distance from the center C, and would be entirely independent of the external shell, as ED.

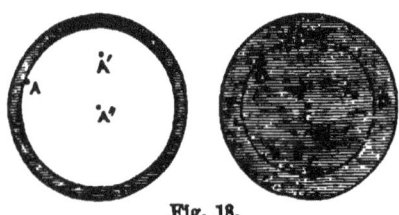

Fig. 18.

P. 25. In performing Exp. 1 notice the difference in the crystals formed on the thread and those formed on the bottom of the vessel. The latter are said to be *tabulated*.

In performing Exp. 2 watch the growth of the crystals, looking through a simple microscope. In their note-books pupils may draw figures of the crystals formed, both of a single crystal, and of a group of crystals, selecting the most interesting.

CHAPTER I. — PROPERTIES OF MATTER. 135

§ 25, p. 26. Repeat the experiment with the bar-magnet and floats (p. 133), and notice that when the floats are attracted to one end of the magnet, through the influence of their polarity, they arrange themselves in some regular geometrical form, *i.e.*, in squares, pentagons, hexagons, etc. This phenomenon is as significant as it is interesting.

§ 28, p. 31. *Strain.* In physics any alteration of size or shape whatever is called a *strain.* It includes all alterations in volume, — as compression or expansion of gases, — all twistings and bendings, and all extensions, as a piece of stretched india-rubber.

§ 30, p. 31. *Viscosity.* " If a constant stress causes a strain or displacement in the body, which increases continually with the time, the substance is said to be *viscous.*

"When this continuous alteration of form is only produced by stresses exceeding a certain value, the substance is called a solid, however soft it may be. When the very smallest stress, if continued long enough, will cause a constantly increasing change of form, the body must be regarded as a *viscous fluid*, however hard it may be.

" Thus, a tallow candle is much softer than a stick of sealing-wax ; but if the candle and the stick of sealing-wax are laid horizontally between two supports, the sealing-wax will in a few weeks in summer bend with its own weight, while the candle remains straight. The candle is therefore a soft solid, and the sealing-wax a very viscous fluid.

" What is required to alter the form of a soft solid is a sufficient force ; and this, when applied, produces its effect at once. In the case of a viscous fluid, it is *time* which is required ; and, if enough time is given, the very smallest force will produce a sensible effect, such as would require a very large force if suddenly applied." — MAXWELL.

§ 33, p. 33. *Adhesion.* In both experiments with the water and the mercury it is important that the glass slip should be quite clean. It is well, previous to each experiment, to wipe the glass and the surface of the mercury with a dry, clean cloth, so as to remove any oxide of mercury which may rest upon their surfaces.

CHAPTER II.

DYNAMICS.

P. 47. If the author's "Seven-in-one" apparatus is substituted for the Magdeburg hemispheres, no air-pump will be needed, and no mystery, which its action and the necessity for removal of air might cause, will be introduced. If the piston of this apparatus is forced into the end of the cylinder, and the stop-cock is closed so as to prevent air entering the apparatus, a weight of two or three hundred pounds may be suspended from the piston without drawing it down. But if air is admitted, by turning the stop-cock so as to press on both sides of the piston, it will quickly descend.

§ 46, p. 48. *Apparatus for exploration of pressure in the interior of a liquid mass.* A very convenient substitute for the apparatus represented in Fig. 27 is a glass manometer tube, represented in Fig. 19. Mercury or some lighter colored liquid may be used in the tube: the lighter the liquid the more sensitive will be the instrument. Connecting a short rubber tube with the extremity *a* of the glass tube, and bending it in different directions, pressure in all directions and at different depths can be explored and compared.

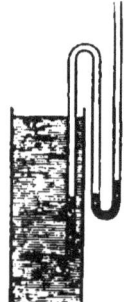

Fig. 19.

§ 50, p. 57. *Mariotte's Law Apparatus.* In the preparation of this apparatus for use there is usually some difficulty in getting the surfaces of the mercury in the two arms of the tube on the same level. This may be accomplished, however, after a few trials, by tipping the tube, and either admitting small bubbles of air into the short arm, or excluding it therefrom, as the case may require.

The base of the tube (Fig. 38) is sufficiently large to receive the tube (Fig. 39), hence it will answer for the jar B in Exp. 2.

§ 52, p. 61. In using the apparatus illustrated in Fig. 46, very likely the stream issuing from the longest tube d may not quite reach the level of the top of the other streams. This is to be explained as the result of the friction against the sides of the longer tube, the friction increasing with the length of the tube.

§ 53, p. 64. In using the Seven-in-one apparatus as a hydrostatic bellows, the piston should be forced into the cylinder, and the space between the piston and the end of the cylinder should be filled with water. This may be done by temporarily removing the rubber tube. If there is any difficulty in removing the air from the tube, so as to allow the water to enter it, the tube may be first filled with water, and, while in a nearly horizontal position, be connected with the union-screw, and afterwards raised to a vertical position.

§ 55, p. 67. In using the improved Pascal's Vases, support the base upon the side of a water pail. Attach to it first the vase corresponding to C, Fig. 52. Suspend the disk d from one arm of the balance-beam, and the counterpoise from one of the holes in the other arm. Pour water slowly into the vase, allowing it to trickle down its side, at the same time elevating the nut on the rod supporting the disk so as to keep it on a level with the surface of the water. Continue to pour water until it forces the bottom off. Then remove the vase from the base and substitute the cylinder A. Now, if water is carefully poured into the cylinder, it will be found that the bottom will be forced off at the instant the surface of the water reaches the nut.

§ 63, p. 80. *Specific Gravity.* One of the pans of the balances furnished by the author has a hook beneath it, from which specimens are suspended. The gram weights accompanying the same are made of brass, and the centimeter and millimeter weights of aluminum; and the whole are neatly mounted in a block of wood.

The pupil will soon discover (however simple it may seem to him before trial) that weighing is an art at which he will find himself quite awkward at first. He will learn it better by experience than precept. A few directions, like the following, will be serviceable to him.

Always ascertain the weight of a solid in air, before it is wet by the liquid.

When weighing in a liquid, see that the solid is completely immersed, and nowhere touches the vessel holding the liquid.

While changing the weights, hold the beam with one hand, that it may not fall from its support and suffer injury.

It is not desirable to use specimens weighing in the air more than from 5^g to 8^g. Specimens even lighter than these will answer just as well. After weighing one or two specimens, the pupil will progress very rapidly, will be delighted with the work, and should be allowed to use as many specimens as time will permit.

§ 74, p. 98. This experiment is hardly practicable, but its description will serve to indicate to the pupil the *true* method of finding the centre of gravity of a mass.

§ 77, p. 104. The plank, Fig. 86, should have a shallow groove cut in it, to guide the ball. Teachers report very satisfactory results from this experiment.

§ 81, p. 108. "I require my pupils to draw the paths of projectiles at different angles, as indicated by the streams of water from the Eight-in-one apparatus with the tube elevated at different angles, and have obtained good results." — G. F. FORBES, Roxbury Latin School, Boston.

§ 81, p. 110. *Improved Apparatus for verifying the Second Law of Motion.* The rod d, Fig. 20, is drawn back toward the left, and the detent pin c is placed in one of the three slots. One of the brass balls is then placed on the projecting rod, and in contact with the end of the instrument as at A; the other ball is placed in the tube B. Release the detent, and the ball at B, struck by the rod d, is projected with a force (when the

CHAPTER II. — DYNAMICS. 139

spring is under its greatest strain) of about 15 lbs. At the instant it escapes the end of tube B, and is free to fall, the rod leaves ball A, and the latter begins to fall. Both balls are pierced with holes, in order that the masses of both may be equal. It is believed that this apparatus is the only one of the kind which has given entire satisfaction.

Fig. 91, p. 111. For the pendulum A six iron balls are used, pierced by holes through their centers, through which a string is passed. The balls are kept in place by knots tied in the string.

The length of a pendulum, e.g., B, is approximately the distance from the point of suspension to the center of the ball.

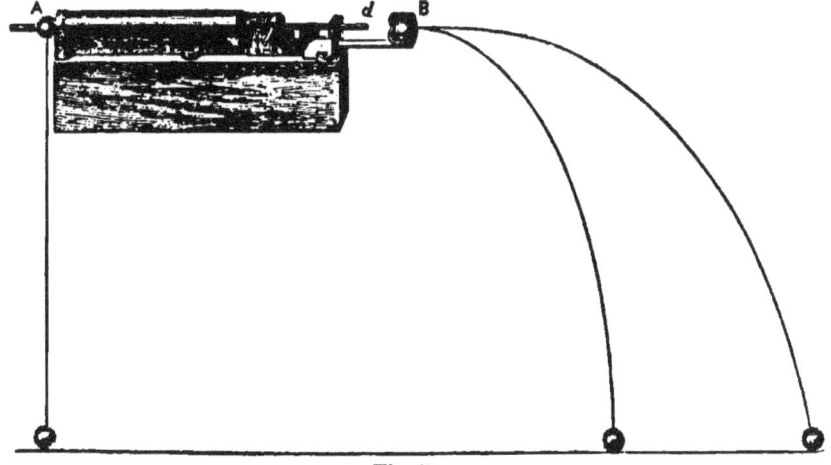

Fig. 20.

§ 82, p. 111. The center of oscillation is that point of a pendulum that vibrates in the same time, as if free from the influence of all other particles.

§ 88, p. 119. The usual mechanical definition of work is here given. A complete dynamical definition, of course, should be made to include negative as well as positive work, and may be stated as follows: When a force moves a body against resistance, or *alters the rate of motion of a body*, it is said to do work.

§ 92, p. 121. Potential energy — energy of position — has been called energy of stress, *i.e.*, energy due to stress. It is really due to both position and stress. If we define stress provisionally as a pressure or a pull, then we shall find that pressure is transformed into the energy of motion *if the bodies pressed upon can move*, *i.e.*, are *in a position* which admits of motion.

§ 96, p. 126. The erg may also be defined as the work done in the latitude of the Northern States by raising $\frac{1}{980}$ of a gram-mass to the hight of 1^{cm}. The objection to this method of definition is that it is a variable measure. The same may be said of the foot-pound, which also evidently varies with locality, and must be reduced at each place to absolute units by the equation

$$\text{Work} = Fs = \text{weight} \times s = Mgs.$$

For example, in the Northern States the work done in raising a pound-mass through 1 foot is

$$Mgs = 1 \times 32.2 \times 1 = 32.2 \text{ foot-poundals.}$$

§ 101, p. 131. *Power.* The pupil should be informed that, in connection with machines, the term *power* has a technical signification which is quite distinct from its usual signification in dynamics. Here it is used in the sense of *force*.

§ 101, p. 132. *Law of Machines.* In a perfect machine (*i.e.*, one in which no internal work is done) the work done upon the machine ($= Fs$) is equal to the work done by the machine ($= F_1 s_1$); or, $Fs = F_1 s_1$; hence, $F : F_1 :: s : s_1$.

P. 135. *Levers.* If a teacher has a "plenty of time" (?) he may teach the popular but useless division of levers into three classes.

CHAPTER III.

HEAT.

§ 103, p. 139. When the water in this experiment nearly reaches the boiling-point, it will be forced out in a constant stream. Previous to that, it will only escape in drops.

§ 110, p. 142, Exp. 4. In the apparatus prepared especially for this experiment, several wires of different metals are fastened

CHAPTER III. — HEAT.

upon a board, and made to converge to a point where the flame is to be located. By running the fingers along the several wires toward the heated end, until they reach the point in each where the heat is unendurable, and noting the distance of these points respectively from the flame, the pupil is enabled to determine with a great degree of accuracy the relative conductivities of the several metals. He will notice that those metals which are the poorest conductors become incandescent first, and should be required to explain this phenomenon.

Exp. 5. The pupil should be directed carefully to avoid allowing the flame to touch the part of the tube not covered by the water.

§ 111, Exp. 1. The success of this experiment will depend largely upon the skill in manipulation, and had better be performed by the teacher only. It will be well to heat the water in the beaker as warm as can be borne by the hands before inserting the tube.

§ 112, p. 145. *Ventilation.* "The volume of air to be renewed in places requiring to be purified may be fixed as follows : —

Per hour and per individual.

"Hospitals:	
Ordinary cases	60 to 70^{cbm}
Wounded persons	100
At times of epidemics	150
Prisons	50
Workshops, ordinary	60
" unhealthy	100
Theatres, music-halls, etc.	40
Long gatherings, meetings	60
Short gatherings, meetings	30
Infant schools	12 to 15
Adult schools	25 to 30
Stables, etc.	180 to 200

"These figures agree with those of English, American, and German hygienists." — GENERAL MORIN, Director of the Conservatoire des Arts et Métiérs, at Paris.

§ 127, p. 158. "Let us suppose that the rarefaction is carried on so far that only one particle out of every original million is left in the space exhausted. The pressure is one-millionth of its original amount; but any molecule once in motion has one-millionth its former chance of encountering any other molecule, and, consequently, its average free-path is magnified a millionfold. The mean path would then be (Crookes) $\frac{1}{10000}^{mm}$ × 1,000,000 = 100^{mm}, or about 4 inches. By means of a good Sprengel pump, exhaustion to the hundred-millionth of an atmosphere can be attained, and the mean free-path of the gas so rarefied would be about 33 feet. In our atmosphere, at a hight of 210 miles, the single molecules are relatively so few (1000 to the ccm.) that each molecule might travel through a uniform atmosphere of that density for 60,000,000 miles without entering into collision. Beyond a hight of 300 miles, the atmosphere is so rare (less than one molecule per cubic foot) that the particles might freely travel through such an atmosphere from one fixed star to another." — DANIELL.

§ 116, p. 150. Other examples of abnormal expansion and contraction are Rose's fusible metal, iodide of silver, and India rubber. See "Elementary Treatise on Heat," by Balfour Stewart, 4th ed., p. 40.

P. 159, Exp. 3. Inasmuch as the water first receives the heat, and then communicates it to the ice, it is not possible to prevent the water, even though we constantly stir it, from becoming warmer than the ice. But the temperature of the ice will not rise above 0° C.

P. 161. *Boiling point.* "Dr. Carnelly finds that ice, if heated under an exceedingly small pressure, may be rendered very hot (180° C.), and will volatilize freely, yet without melting, unless the pressure be allowed to exceed a certain low maximum, which he calls. the *critical pressure.*" — DANIELL.

§ 129, p. 162. The condenser furnished by the author consists of a vessel having a capacity of four to five liters, in which is coiled six feet of pure copper tube. The glass delivery tube b, Fig. 115, is connected with one end of this copper tube by a rubber connector. From the other end of the copper tube which pierces the vessel, near its bottom, escapes the distilled liquid. Cold water is siphoned into the condenser, as in the figure, and the heated water escapes near the top of the vessel through a delivery tube into a sink.

§ 130, p. 163. *Molecular Theory of Evaporation and Condensation.* "We have seen that in the case of a gas, some of the molecules have a much greater velocity than others, so that it is only to the average velocity of all the molecules that we can ascribe a definite value. It is probable that this is also true of the motions of the molecules of liquids, so that, though the average velocity may be much smaller than in the vapor of that liquid, some of the molecules in the liquid may have velocities equal to or greater than the average velocity in the vapor. If any of the molecules at the surface of the liquid have such velocities, and if they are moving *from* the liquid, they will escape from those forces which retain the other molecules as constituents of the liquid, and will fly about as vapor in the space outside the liquid. This is the *molecular theory of evaporation*. At the same time, a molecule of the vapor striking the liquid may become entangled among the molecules of the liquid, and may thus become part of the liquid. This is the molecular explanation of condensation. The number of molecules which pass from the liquid to the vapor depends on the temperature of the liquid. The number of molecules which pass from the vapor to the liquid depends upon the density of the vapor as well as its temperature. If the temperature of the vapor is the same as that of the liquid, evaporation will take place as long as more molecules are evaporated than condensed; but when the density of the vapor has increased to such a value that as many molecules are condensed as evaporated, then the vapor has attained its maximum density. It is then said to be saturated, and it is commonly supposed that evaporation ceases. According to the molecular theory, however, evaporation is still going on as fast as ever; only, condensation is also going on at an equal rate, since the proportions of liquid and of gas remain unchanged." — MAXWELL.

§ 132, p. 165, Exp. 1. For obvious reasons the results here stated are theoretical rather than practical, inasmuch as the water resulting from the melting ice cannot be kept at the same temperature as the ice.

More satisfactory results may be obtained by pouring 1^k of water at 80° C. upon 1^k of ice at 0° C., and noting the temperature of the liquid at the instant the ice becomes melted, and calculating from the data found the number of units of heat rendered latent.

§ 133, p. 167. *Latent heat.* "We now know that it is not heat of any kind; it is latent or potential energy. Work must be done upon ice in order to convert it into the more highly-stressed condition of water. Water differs from ice at the same temperature in possessing more potential energy." — DANIELL.

§ 135, p. 167. The reason for using two substances in this experiment rather than one is that a given quantity of water will dissolve more of both than of either alone; consequently, a greater amount of heat will be consumed.

§ 136, p. 167, Exp. 2. For the purpose of comparison, it would be well to take water at about 60° C., and fill the porous cup, and also a glass beaker of as nearly the same size and shape as practicable, and place a thermometer in each. In the course of five to ten minutes there will be quite a perceptible change in the temperature of the two bodies of liquid which had the same temperature at the beginning.

EXP. 3. The author finds that raising a window a little way, so as to get a good draft of air, answers much better than the use of the bellows. It may take at best from ten to fifteen minutes to freeze the water. Hence, it would be well to reduce

CHAPTER III. — HEAT.

the temperature of the water to a low point by means of a freezing mixture before introducing it into the tube.

§ 145, p. 174. The phrase "conservation of force" is sometimes used, but should be avoided, because it is entirely erroneous. A single illustration will make this apparent. Take a lever 12 ft. long, place the fulcrum 3 ft. from one end, and apply a force of 3 lbs. at the extremity of the long arm, and a force of 9 lbs. will be exerted at the other end of the lever. Here force is apparently created. If the force is applied at the extremity of the short arm, force apparently disappears. But though there is no conservation of force, there is a strict conservation of energy in this and in all other mechanical contrivances. On this principle is to be explained the paradox in the statement that a single-inch piston of the hydraulic press, pressed in with a force of 60 lbs., will commensurate a pressure of 60 lbs. to every square inch of a cylinder, however large.

You can store energy, but you cannot store force any more than you can store time. A stone resting on the ground presses the ground: force is all the time exerted, but the cleverest engineer could not drive a machine by using a weight resting on the ground.

§ 147, p. 175. *Joule's Equivalent.* According to Joule's revision of this physical constant, its value is, at sea-level at the latitude of Greenwich, 772.55 ft. lbs. In accordance with this value, the calorie = 423.985^{kgm} = 41,593,010,000 ergs.

If the numerical value is so chosen as to give the work corresponding to a unit of heat, it is called *Joule's Equivalent*, or the mechanical equivalent of heat; if, on the contrary, it gives the heat corresponding to a unit of work, it is called *the thermal equivalent of work.* If the former is denoted by J, the latter is $\frac{1}{J} = \frac{1}{423.985} = 0.00235$ caloric; *i.e.*, 1^{kgm} of work is equivalent to 0.00235 calorie of heat.

CHAPTER IV.

ELECTRICITY.

§ 151, p. 181. *Current.* Electricity, whatever it is, can pass from one body to another only by passing consecutively through every point of the path joining them. We may, therefore, with perfect propriety speak of a *current* of electricity.

§ 158, p. 184. "*How Electricity Originates.*" In the first edition of the Physics the foregoing expression, as well as the expression "generate electricity," were inadvertently used. These expressions, though often convenient, ought carefully to be avoided, as they convey erroneous ideas. Electricity is a *something* whose sum total in the universe seems to be constant, for we cannot alter the quantity contained in an isolated space by any method. We conclude that electricity is indestructible and uncreatable, by an exact parity of reasoning with that by which we are convinced that matter is indestructible and uncreatable.

When a body is electrified, another is always charged with an equal amount of the opposite kind of electricity, so that we may regard the process as, not the generation of anything, but a *separation*.

Exp. 4, p. 194. A suitable battery for this experiment consists of two Bunsen cells connected tandem. The electrolyte may be composed of sulphuric acid diluted with twenty times its volume of water.

§ 178, p. 203. "The resistance of all wires increases as the temperature rises, and the resistance of nearly all metals increases at the same rate, iron and thallium, according to Dr. Matthiesen, being the only exceptions. From the tables given by Latimer Clark we learn that the resistance of iron wire increases about thirty-five hundredths (0.35) per cent for each degree Fahrenheit, and that the resistance of copper increases, as the temperature rises, twenty-one hundredths (0.21) per cent for each degree.

"The rate of increase is not reckoned all through on the original resistance, but is computed in the same manner as compound interest on a sum of money. For example, if we have a wire which measures 100 ohms at 60° F., and the resistance be increased a certain amount by a rise of one degree in temperature, it will be increased by the next degree of rise at the same rate per cent, calculated on the original resistance, plus the amount increased by the first degree of rise." — LOCKWOOD.

Exp. 1, p. 215. As many as four Bunsen cells, connected abreast, should be used in this experiment, and the extremities of the wires should not dip more than 1^{mm} into the mercury. In using the apparatus illustrated in Fig. 152, a very strong current will be required.

Exp. 2, p. 216. In the apparatus furnished by the author for this experiment, zinc and carbon plates are used, and a solution of bichromate of potash, like that used in a Grenet cell, should be used in this floating battery. The battery, left to itself, will take up a position with its coil N and S.

§ 190, p. 218. Ampère's theory serves a very useful purpose in acquainting the pupil with very many phenomena of electricity and magnetism, and the laws governing them, very much as the fluid theory of electricity has, at least, furnished a very convenient language in which to express the various electrical phenomena. Yet, plausible as this theory seems in many of its aspects, it is open to many serious objections, of which we give only one. It would seem that, if this theory be true, force would be required to *demagnetize* instead of to magnetize matter; for, if the assumed molecular currents forming an inherent part of the constitution of matter really exist, they must, by the fundamental laws of electric currents, always arrange themselves in parallel order.

Exp. 3, p. 221. In performing this experiment, the glass plate should be thin, and should not touch the steel disk. It would be well to have a glass plate set in a wooden frame for this purpose.

§ 192, p. 222. *Magnetic Poles of the Earth, and Variation of the Needle.* "When the phenomena of terrestrial magnetism were first somewhat accurately observed, about 300 years ago, the needle pointed here in England a little to the east of north. A few years later it pointed due north; then, until about the year 1820, it went to the west of north, and now it is coming back towards the north. . . . Everything goes on as if the earth had a magnetic pole revolving at a distance of about twenty degrees round the true North Pole. . . . About 200 years from now we may expect the magnetic pole to be between England and the North Pole; and in England at that time the needle will point due north, and the dip will be greater than it has been for 1000 years, or will be again for another. That motion of the magnetic pole in a circle round the true North Pole has already, within the period during which accurate measurements have been made, extended to somewhat more than a quarter of the whole revolution." — Sir William Thomson.

§ 207, p. 237. The principal source of difference of potential is the *contact of dissimilar surfaces,* — that is, either of different substances, or of two pieces of the same substance whose surfaces are in different conditions. A piece of rosin and a piece of glass will, after contact, be more difficult to pull asunder than two pieces of rosin or two pieces of glass; and if they be rubbed together, so as to multiply the points of contact, the effect is multiplied. When pulled asunder, two such bodies are found to be charged equally and oppositely: across the surface of contact there has been a separation of positive from negative electricity. The development of electrical condition is thus necessarily a phenomenon of continual recurrence, and it greatly influences the adhesion of one body to another. In all probability, wherever there is friction, the energy ultimately converted into heat is, in the first place, converted into the energy of electrical separation.

CHAPTER IV. — ELECTRICITY.

"When two substances have different molecular velocities at their common surface of mutual contact, the molecules hamper one another, and energy is lost: this energy takes the form of the energy of electrical displacement." — DANIELL.

§ 233, p. 250. The quickest way to charge a battery of jars is to connect the inner coatings with one of the conductors of the electrical machine, and the outer coatings with the other conductor. Likewise, in using the Aurora tube, the top of the tube should be connected with one conductor and the bottom of the tube with the other. And the same is true of all other pieces of apparatus through which charges of electricity are to be sent.

Fig. 192, p. 254. A good substitute for the apparatus here described may be easily and cheaply prepared as follows: Apply a varnish, made by dissolving gum shellac in alcohol, to one side of a piece of window-glass about 6 inches long and 4 inches wide, and sift iron filings over the wet surface. As the alcohol evaporates, the gum will cause the filings to adhere firmly to the glass. The glass thus prepared may be used in the manner directed for the mica disk.

A piece of common mirror-glass also answers this purpose well.

§ 230, p. 256. Great E.M.F. is produced in every frictional machine, but it is an essential part of the contrivance that the rubber and the main conductor shall be separated by the insulator which is rubbed. We cannot, therefore, even if we connect the rubbers with the main conductor, complete a circuit, except through this insulator. Now this insulator has such an enormous resistance that, according to Ohm's Law, even the great E.M.F. produced by the machine can produce only a very feeble current.

§ 234, p. 259. *Voltaic Arc.* One may form a good idea of the size and shape of the arc by looking at it through a colored or smoked glass.

§ 245, p. 269. The following very clear and concise description of the operation of the telephone is taken, with slight modification, from Jenkins's "Manual of Electricity":—

"One instrument, which shall be called the receiver, is held to the ear. The person wishing to send a message speaks into the mouthpiece of the other instrument, which will be called the sender. The spoken words cause the air to vibrate in front of the sender, and the disk E of that instrument vibrates as the air does, alternately approaching and leaving the end of the magnet M. Each change in the position of the disk E causes a change in its magnetism, and in the magnetic field occupied by the coil B. Each change in the magnetic field causes an induced current in the circuit. This current is reversed at each change of direction in the motion of the disk, and, moreover, its magnitude is at each instant sensibly proportional to the rate at which the disk E is moving; for we know that the induced current is proportional to the rate of change in the field enclosed by the coil B, and we see that this rate of change will depend on the rate at which the disk E is moving. The induced currents acting on the receiving instrument will change the magnetism of the steel magnet A and the magnetic field in which the disk E of the receiver lies; each change will be accompanied by a change in the attraction of the iron disk to the magnet, and thus the disk E will be set in vibration. It will move to and fro as often as the direction of the current in the circuit is reversed; but, more than this, its rate of motion at each instant will be proportional to the rate of change in the magnetic field it occupies. Now, this rate of change is the same rate as that of the change in the current, which again is the same rate as that of the motion of the sending disk E. The motions of the sending and receiving disks will therefore be similar, though of unequal magnitude. The air, therefore, in front of them — which in one case moves the disk, and in the other is moved by it — will also vibrate in the same way; and, since the vibrations of the air at the sending end produced the impression of articulate words on the ear, so the vibrations of the air caused by the disk *E* at the receiving end will also produce the impression of the same articulate sounds. The chief difference between the two sounds is one of magnitude. The action of the two disks is similar to that in the toy telegraph, where two parchment disks are mechanically connected by a tight string. The electrical currents due to induction give those impulses in the one case which in the other are given mechanically by the string."

CHAPTER V.

SOUND.

§ 267, p. 288. *Loudness of Sound.* "The sensible loudness of sounds does not coincide very closely with their physical intensity. This arises partly from modification in the form of the vibration induced by so complicated a transmission through the auditory apparatus, partly from causes purely physiological."— DANIELL.

§ 274, p. 294. Very fair results may be obtained in these experiments with an ordinary tuning-fork. The forks mounted on resonance-boxes are considerably larger and more expensive than the ordinary tuning-fork, and are usually called *diapasons*. They should be set in vibration by bowing with a large bass bow, the bow having been previously rubbed over a piece of warm rosin. Avoid striking them upon hard substances, as permanent injury may be done them in this way. Another convenient way of setting them in vibration is by drawing quickly between the tines a rod of wood whose diameter is a little greater than the width of the space between the tines at their extremities. Diapasons should be carefully protected from rust, as this will alter their pitch. Diapasons like those in Fig. 214 require especial care, as a very little rust, or a very slight change in the elasticity of either one, may alter by a single vibration, or a part of a vibration, the vibration rate of one of them. In that case they must be tuned in unison by an experienced hand before they will answer again the purpose of showing sympathetic vibrations. To protect forks from rust, previous to laying away after use, they should be wiped with a woollen cloth slightly moistened with vaseline. Vaseline may be used to advantage to protect all pieces of apparatus made of iron or steel from rust.

§ 296, p. 321. The lowest sound is obtained by touching the center of one side and bowing the corner. The damping is best done by touching the plate with the extremity of the finger-nail. A node is always started from the point that is touched, while, of course, the point bowed is a ventral segment. The next note, a fifth above, is produced by damping the corner and bowing the center. By altering the position of the finger and bow, and sometimes using finger and thumb, a great variety of figures may be obtained, which may be further extended by changing the points of support.

If sand (fine writing sand is best) mixed with lycopodium powder is strewed upon a vibrating plate, the sand will collect on the nodal lines; but the lycopodium, by the agitation of the air, will be *blown* toward the center of each vibrating segment.

CHAPTER VI.

LIGHT.

§ 300, p. 325. *Kinds of Radiation.* "When a succession of waves impinges on a mass of ordinary matter, the effect varies according to the nature and the condition of the body which receives their shock; if it be an ordinary opaque mass, that mass may be warmed, wave-motion being transformed into heat, and the waves, which have impinged upon it, are *ex post facto* called a beam of radiant heat; if they fall upon the eye, they may produce a sensation of light, and the wave system is then called a beam of light; falling upon a sensitized photographic plate, or a living green leaf, it may operate chemical decomposition, and it is then called a beam of actinic rays. The word "rays" in the last phrase may be understood to mean, not imaginary lines at right angles to the wave-front, but kinds of radiation; and hence we speak of heat rays, of light rays, of chemical or actinic rays, these names being given to one and the same train of waves according to the effects which it is found competent to produce. But while ether-waves are in course of traversing the ether, there is neither heat, light, nor chemical decomposition; merely wave-motion and transference of energy by wave-motion. Hence, none of these names can in strictness be applied to a train of waves while these are actually travelling through the ether.

"According to Clerk Maxwell's view, the ether is a homogeneous body, a non-conductor of electricity. Periodic electric-stresses applied to this produce waves which travel at the rate of about 300,000,000 meters per second. These waves are waves of transverse vibration, and there is no vibration longitudinal or normal to the wave-front." — DANIELL.

§ 311, p. 334. In a course of lectures given at the Lowell Institute in Boston, in 1882, Prof. S. P. Langley said that the light of the sun is two and a half times as brilliant as the same area of electric (arc) light; and that if a calcium light be held between the eye and the sun, the light would appear to be a black spot upon the sun. As a measure of comparison, in assisting the comprehension of the infinite quantity of light thrown off by the sun, he remarked that if there was an electric light of 2000 candle power on each square foot of the surface of the earth, then the whole light from the earth would be less than one-billionth that from the sun.

§ 301, p. 326. "We are led to infer, therefore, that there is such a medium, which we call the *luminiferous ether*, or simply the *ether;* that it can convey energy; that it can present it at any instant, partly in the form of kinetic, partly in that of potential energy; that it is, therefore, capable of displacement and of tension; and that it must have rigidity and elasticity. Calculation leads us to infer that its density is (Clerk Maxwell) $\frac{936}{1,000,000,000,000,000,000,000}$ that of water, or equal to that of our atmosphere at a hight of about 210 miles, a density vastly greater than that of the same atmosphere in the interstellar spaces; that its rigidity is about $\frac{1}{1,000,000,000}$ that of steel; hence, that it is easily displaceable by a moving mass; that it is not discontinuous or granular; and hence, that, as a whole, it may be compared to an impalpable and all-pervading jelly, through which *light* and *heat waves* are constantly throbbing, which is constantly being set in local strains and released from them, and being whirled in local vortices, thus producing the various phenomena of electricity and magnetism; and through which the particles of ordinary matter move freely, encountering but little retardation if any; for its elasticity, as it closes up behind each moving particle, is approximately perfect." — DANIELL.

"We are at liberty to deny the existence of all action at a distance, and attribute it to the intervening medium, which, to be logical, we must assume to be continuous and not molecular in constitution." — ROWLAND.

"It is a most wonderful fact that we have never been able to discover anything on the earth by which our motion through a medium can be directly proved. Carried, as we suppose, by the earth with immense velocity through the regions of space filled with ether, we have never yet been able to prove any direct influence from this ethereal wind." — ROWLAND.

§ 344, p. 373. "The bolometer, a curious instrument recently introduced by Prof. S. P. Langley for measuring minute quantities of radiant energy, promises important results in optical and astronomical investigations. It is based upon the fact that, when equal conductors of the electrical current are at the same temperature, their conductivities are equal, and the current of a battery can be equally divided between them; while, if unequally heated, their conductivities are unequal, and the difference in current can be detected with the galvanometer. By substituting thin sheets of metal for the wires ordinarily employed as conductors, so as to take up and part with its radiations with great rapidity, an instrument is produced capable of measuring such minute quantities of heat as $0.00001°$ C.; capable, also, of recording the infinitesimal heat radiations of the diffraction spectrum. The interesting statement is made in this connection, that the curves of light, heat, and actinism, instead of receding from each other, as commonly understood, are in reality coincident, that is, *the solar beam, instead of consisting of a pencil of rays bound into a luminous sheaf called light, is a homogeneous and simple energy, the names light, heat, and actinism being merely names for its different modes.*" — ELECTRICIAN.

§ 346, p. 375. *Effect of the Atmosphere on the Color of Sunlight.* "Sunlight is originally bright blue, and is extremely rich in the more refrangible rays, but filtration through two absorbent atmospheres — that of the sun and that of the earth — renders it a yellowish white." — LANGLEY.

§ 347, p. 378. It is now the received view that all color-perceptions, infinite as they may be in intensity and in hue, are due to the simultaneous excitation of three sets of nerve-ends by stimuli of relatively varying amount. These three physiologically-primary color sensations are (Young and Helmholtz) red, green, and violet. When orange light affects the eye, the nerve-ends sensitive to red are affected; those sensitive to green are affected, but less so, while those sensitive to violet are very feebly affected. When the red and green nerve-ends (as we may for convenience call them) are equally affected, the resultant impression is one of yellow; hence, a mixture of red *light* with green *light* produces a sensation of yellow. In like manner, the mixture of green and violet in varying proportions may produce all the color-sensations which the spectrum between the green and violet is capable of stimulating, as may be shown by the rotation of suitable color-disks.

§ 351, p. 379. *Interference.* "The twinkling of stars is another effect of interference: light coming to the eye from a star, so distant as to be practically a single luminous point, arrives in rays which have traversed slightly unequal distances in an irregularly refracting atmosphere, and thus enter the eye in unequal phases. Now one color is distinguished, now another; the eye perceives colored light complementary to that momentarily lost. No two persons can, as a rule, see any star twinkling in precisely the same manner. The planets twinkle only at their edges; their disks present many points or sources of light, whose scintillations, on the whole, mask one another." — DANIELL.

§ 353, p. 384. *Polarization.* "Ordinary light consists of vibrations taking place always in planes at right angles to the direction of the ray, but in all directions in those planes. That is, if the ray travels along the axle of a wheel, the vibrations composing it are all in the plane of the wheel, but are executed along any or all of the spokes.

"The effect of reflecting light at certain angles from certain substances, or of passing it through certain crystalline substances, is to cause all the vibrations to take place in the same direction, — that is, along one spoke of the wheel and the spoke opposite to it.

"The light is then said to be polarized. Now, if the wheel, without being rotated, be slid along the axle, the spoke along which the vibrations take place will trace out a plane.

"When no rotative force is applied to the polarized light, the vibrations all take place in this plane, and the light is said to be 'plane-polarized.'

"We cannot detect by the eye in what plane light is polarized, or, indeed, whether or not it is polarized at all. In order to do so, we have to take advantage of the following natural law: —

"Transparent bodies which have the power of polarizing light in any given plane are opaque to light already polarized in a plane at right angles to that plane; and reflecting surfaces which have the power of polarizing light in a given plane will not reflect light which, when it falls on them, is already polarized in a plane at right angles to that plane.

"Thus, to determine in what plane light is polarized, we have only to take a crystal which has the power of polarizing light in a certain plane, fixed with regard to its axis, and to turn it round till the light is extinguished.

"We then know that the light is polarized in a plane at right angles to that plane in the crystal." — GORDON.

§ 356, p. 387. "Whatever light is, at each point of space there is something going on, whether displacement, or rotation, or something not yet imagined, but which is certainly of the nature of a directed quantity, the direction of which is normal to the direction of the ray. This is completely proved by the phenomena of interference." — MAXWELL.

Part IV.

TEST QUESTIONS.

1. Describe an experiment which you have performed at your home, and state the lesson derived from it.

2. Why, as you raise the vessel B, Fig. 22, Physics, higher and higher, is the rubber forced inward more and more?

3. Name some phenomena which are the result of the earth's attracting the moon and the moon's attracting the earth. — *Ans.* The former attraction mainly keeps the moon in her orbit, and the latter is one of the causes of tidal phenomena.

4. What is the distinction between *mass* and *weight?* Which better defines a body? Why?

5. If the earth were a homogeneous shell devoid of air, and a person were to jump from one side toward the center, where would he stop? Could he stop at the center? Would his path be straight or curved? Would his motion be accelerated, retarded, or uniform? What would be the effect produced upon the earth at the instant he jumps? If the earth and his body were perfectly elastic (*i.e.*, the coefficient of restitution $= 1$), how long would he continue to move? If in his journey through the hollow space he should let drop a ball, what would become of it?

6. If the earth were a homogeneous sphere, and a hole extended from surface to surface through the center, and the hole were a vacuum, and a ball should be dropped into it, where would it stop? Where would it have its maximum velocity? How long would it continue to move?

TEST QUESTIONS.

7. A coil of glass tubing, after being suspended for a year, became permanently stretched. What property does this phenomenon show that glass possesses?

8. One man holds one end of a rope in his hands, and another man pulls the other end of the rope with a force of 90 lbs. What force does the latter compel the former to exert in order to retain the rope in his hands? — *Ans.* 90 lbs.

9. Two men at opposite extremities of a rope pull each with a force of 100 lbs. What is the force exerted between them or the tension of the rope? — *Ans.* 100 lbs.

10. What force is necessary to separate a pair of Magdeburg hemispheres from which the air has been entirely exhausted, and whose diameter is five inches? — *Ans.* 294.5 lbs.

11. What force would be necessary to separate the above hemispheres at a place where the barometrical column is 20 inches? — *Ans.* 196.3 lbs.

12. What force would be necessary to separate the same hemispheres at sea-level if only one-fourth of the air has been removed from them? — *Ans.* 73.8 lbs.

13. A stone weighing a kilo rests upon a shelf, and another stone of the same weight is suspended by a string. What effect is produced by the force of gravity acting on each? — *Ans.* In the former case, pressure; in the latter, tension.

14. If the shelf is removed and the string is cut, what change in the effects of gravity will occur? — *Ans.* Pressure and tension will cease, and motion will be produced.

15. In the experiment with vessel B, Fig. 22, why is the rubber pressed in farther the higher the vessel is raised? Is it because the pressure of the air increases as the vessel is raised?

16. A cubical vessel, whose interior dimension is 6^{cm}, is filled with water and sits upon a table. What is the entire pressure exerted by the liquid tending to separate the sides of the containing vessel? — *Ans.* 216^g. The pressure of the liquid upon the bottom is 216^g. But this pressure does not tend to separate the bottom if there is no upward pressure against the top, the

TEST QUESTIONS. 161

weight of the liquid being supported by the table. The pressure against each of the four sides is 108^g, and the force tending to separate two opposite sides is 108^g; and the entire force tending to separate the two pairs of opposite sides is therefore 216^g.

17. A man jumps from an eminence. During his descent how does the pressure of his feet upon the soles of his shoes compare with his weight? Explain.

18. At what hight does a water barometer stand when the mercurial barometer stands at 30 inches? — *Ans.* $33\frac{3}{4}$ ft.

19. In an atmosphere where the pressure is two atmospheres, how long should a barometer tube be to measure the atmospheric pressure?

20. State Mariotte's Law, and how it may be verified.

21. In the Eight-in-one apparatus, why does not water flow from the orifices *b*, *c*, etc., when the plug *a* is out? — *Ans.* The descent of the water is unresisted; consequently, there is no downward pressure, and, for this reason, no lateral pressure.

— 22. If a man slides down a vertical rope, grasping it more or less firmly with a constant grip, how will the tension on the rope compare with his weight?

23. Why is there no lateral pressure in liquids falling freely? — *Ans.* There is no pressure in such a body of liquid in any direction, inasmuch as, according to the supposition, its motion, caused by the force of gravity, is unimpeded. See Physics, p. 44.

24. Place one body on another, and allow them to fall in a vacuum; would the former press upon the latter during the fall?

25. Show that a body having uniform motion must be in a state of equilibrium.

26. Why do fluids transmit pressure in every direction, while solids transmit it usually only in the direction in which the force acts?

27. Why does pressure in a body of liquid increase *as* its depth, while in a body of gas it increases *with* its depth?

28. Lay a piece of paper on the smooth surface of a board, and let both drop. They reach the ground together; but if separated and dropped simultaneously, the board reaches the ground first. Explain.

29. Raise the piston t, Fig. 43, p. 60, to the top of the cylinder s, and stop up the tube u at its opening into the cylinder. What force must be applied to the piston to pull it to the bottom of the cylinder, the area of the transverse section of the piston being 20^{qcm}? Suppose that the piston, at the beginning, is at the middle of the cylinder, will the force required to keep it in motion be constant? About how great will be the force when it reaches the bottom of the barrel? Suppose the force at that point is withdrawn, what will happen? Suppose the apparatus to be inverted, and a person were to blow with a force of 10^g, the area of the cross section of the bore of the tube being 1^{qcm}, what weight placed upon the piston might be sustained? If the free extremity of the tube is raised 2^m above the lower extremity of the piston, and water is poured into the tube until it is filled, what weight placed upon the piston will be sustained by the water? What name would the apparatus receive in the last case? Suppose that a plug, just fitting the interior of the tube, were forced into the tube pressing against the water, what would the apparatus become?

30. The diameter of the mouth of an air-pump receiver is 20^{cm}. Three-fourths of the air has been removed from the receiver. The receiver weighs 1.5^k. What force will be required to raise it from the pump-plate?

31. A person is on deck of a vessel which is moving due east at the rate of one mile an hour; at what rate must he walk due south-west in order that his resultant motion may be due south? What will be his southerly velocity? (Solve by constructing a diagram.)

32. A steamship is moving due north at the rate of 10 miles, the tide carries it due north-east at the rate of 2 miles an hour, while the wind carries it due north-west at the rate of 4 miles an hour; what is its actual course and velocity?

TEST QUESTIONS.

33. A ship is sailing due south-west at the rate of 8 miles an hour; what is its southerly velocity?

34. A boat is crossing a stream at the rate of 5 miles an hour. A person walks from the stern toward the prow at the rate of 3 miles an hour. Describe his several velocities, and state how great they are.

35. While sitting in your chair, what motions has the matter composing your body?

36. A door stands ajar; why is it not moved perceptibly on its hinges when a bullet is fired through it?

37. Draw an oblique line to represent the path of a boat crossing a river, and find the relative intensities of the current, and the force which propels the boat at right angles with the banks.

38. Represent by lines three forces acting at angles with one another on a body, and find their equilibrant.

39. In Fig. 75, p. 93, Physics, what is the relation of the force of gravity acting on the weight w to the forces represented by the lines cA and cB?

40. Locate a point A on your paper, and from it draw a horizontal line AB to the right to represent a force of 10 lbs., acting on a body at A. Draw from A another horizontal line AC to the left to represent another force of 10 lbs., acting on the same body at the same time. In what state will the body be as regards these two forces? What is the relation of each force to the other? — *Ans.* An equilibrant. Show that each force produces its own independent effect in accordance with the second law of motion. Resolve one of the forces into two components. Let the intensities of the two forces be as $8:10$; represent their resultant, and answer the requirement.

41. Draw a vertical line AB. Let this represent the path in which a body at A, the lower extremity of the line, is to move. Draw from A a horizontal line AC to the right, to represent one force acting on the body. Construct a parallelogram, and find the direction and intensity of the other force.

Letter the line which represents it AD. What is the effect produced by the force represented by the line AC? Show that this force produces the same effect that it would produce if the force represented by AD were not acting on the body. In order to do this you must suppose the force acting in the line AD to be resolved into two components. What lines of your parallelogram represent them?

42. Represent by a parallelogram a case in which the intensity of the resultant is less than either of its two components. Explain why it is less.

43. (Fig. 75, Physics.) Draw on the blackboard line CD to represent the equilibrant of the force of gravity on W. Indicate the direction also of the strings CA and CB. Then, with CD as a diagonal, and with two of its sides lying in the direction CA and CB, construct a parallelogram; and, with the intensity of the force represented by CD known, ascertain, by comparing each of the sides lying in the line CA and CB with the line CD, the intensity of each of the component forces. Compare the results with the readings of the dynamometers X and Y.

The intensity of the force CD and one of its components CA being known, find the intensity of the other component. Draw the line CD of any desirable length. Draw a line in the direction CA as indicated by the string, making its length in comparison with CD proportional to the given forces. Complete the parallelogram with CD as a diagonal, and the line lying in the direction CA as one of its sides; and ascertain, by comparing the length of the line lying in the direction CB with the length of the line CD, the intensity of the other component. Verify the result by consulting the reading of the dynamometer Y.

44. Two forces of 20^k and 50^k act at an angle of $60°$; find their equilibrant.

45. If two boats just alike are connected by a rope, and two men, one in each boat, pull on the rope, at what point between

them will they meet? At what point if only one man pulls? Why?

46. State three causes for the variation of gravity on the earth's surface.

47. Can you move without the aid of some other body?

48. Bodies at rest, with respect to the surface of the earth, are really in motion, and their motion is not constant nor in a straight line. Are the forces which act on them in equilibrium?

49. Upon which will the effect of a given force be greater, a body at rest or a body in motion?

50. Express the atmospheric pressure at sea-level in absolute units. — *Ans.* 1,012,634 dynes per square centimeter.

51. Why are "top-heavy" bodies unstable?

52. What mechanical advantage may be gained in a copying press in which the hands move through 1 inch, while the end of the screw descends $\frac{1}{142}$ inch? — *Ans.* $F = 142$.

53. What is the true way of measuring gravity or any other force? — *Ans.* By its effects in producing momentum.

54. How many cubic feet of water will a 10-horse-power engine raise in an hour from a mine 300 feet deep, a cubic ft. of water weighing $62\frac{1}{2}$ lbs.?

55. When a force acts on a body at right angles to the direction of its motion, so as to cause it to revolve in a circle, does it do work on the body? Why?

56. Does the sun do work on the planets, which revolve about it? Explain. — *Ans.* No; the force of its attraction merely alters the direction of their motion, but not their velocities, and, consequently, not their kinetic energy.

57. What is the weight of a body at any place? — *Ans.* It is its mass multiplied by the force of gravity at that place ($W = Mg$).

58. What force is required to lift one gram one centimeter? — *Ans.* 980 dynes in this latitude.

59. (a) A man whose weight is W stands on the platform of an elevator as it descends. If the platform descends with a

uniform acceleration of $\frac{1}{4}^g$, what will be his pressure on the platform? (b) What will it be if the platform ascends with the same uniform acceleration? — *Ans.* (a) $\frac{3}{4}$ *W.* (b) $\frac{5}{4}$ *W.*

60. A stone weighing 15 lbs. lying upon the ground has a spring balance attached to it. A man raises the stone by pulling the spring balance. Will the force employed, as indicated by the spring balance, exceed 15 lbs., and why? If it exceeds, upon what will the excess depend? — *Ans.* It will exceed 15 lbs., the excess being employed in producing motion; and the magnitude of the excess will depend upon the rapidity with which it is moved.

61. A man carrying upon his shoulders a bag of sand weighing 100 lbs., jumps from an eminence. How great will be the pressure of the bag upon his shoulders during the descent, disregarding the resistance of the air? — *Ans.* There will be no pressure, since the man will offer *no resistance*, during the descent, to the action of gravity on the sand.

62. A hammer, whose weight is 1500 lbs., falls 10 ft. How far will it drive a pile into the earth against an average resistance of 10,000 lbs.?

63. What horse-power in a locomotive will be required to draw a train of cars at the rate of 10 miles an hour against a constant resistance of 50 tons?

64. Is friction force? — *Ans.* Yes; since, according to the definition of force, it tends to alter motion.

65. Is work force? — *Ans.* No; work is the product of force multiplied by the space through which it acts.

66. What is the product of force multiplied by the time during which it acts called? — *Ans.* Momentum.

67. When a force acts upon a body and causes it to move a given distance, in what language would you describe the effect of the force? — *Ans.* As work done on the body, or as energy communicated to the body.

68. How does energy differ from power? — *Ans.* The element of time has nothing to do with energy, while power means a capacity to do a given amount of work *in a given time.*

69. If an engine should raise 55 lbs. 10 ft. in a second, and at the end of a second its energy should be exhausted, could it properly be called a one-horse-power engine? — *Ans.* Yes; since while it did work, it performed at the rate of 33,000 ft. lbs. per minute, and this is just as truly a horse-power as it would be if the work were maintained for a thousand years.

70. A cannon ball is shot into empty space; how great a force will be required to deflect it from its path? — *Ans.* Since the body meets with no resistance, any force, however small, will suffice to deflect it from its path in accordance with the Second Law of Motion.

71. Can a child sitting on a sled start or stop the sled by pulling on a cord attached to the sled? Why? — *Ans.* No; since the sled will, in either case, receive both the action and reaction, which, being equal, would neutralize each other.

72. Why does not every body move when acted on by force?

73. Why does a body thrown horizontally into the air fall to the earth?

74. Is the expression " one horse-power per second" admissible, as, for instance, when we wish to convey the idea that a horse-power *lasts*, or is exerted *for* one second? — *Ans.* No; the expression would be equivalent to 33,000 ft. lbs. per second *per second.*

75. How many dynes of force are required to set a mass in motion?

76. How many dynes are required to make a gram-mass move with a velocity of 9.81^m per second, the force acting constantly for one second? What, if it act for two seconds? — *Ans.* 981 dynes; 490.5 dynes.

77. What is the force acting on a falling gram-mass in the Northern States? — *Ans.* 980 dynes.

78. What is the force acting on a falling pound-mass in the Northern States? — *Ans.* 32.191 poundals.

79. How many dynes are required to set a mass weighing 50^k in motion with a velocity of 12^m per second, the force acting for precisely one second? — *Ans.* 60,000,000.

80. What kind of energy is chemical energy? — *Ans.* Potential energy, inasmuch as it is due to forces tending to produce a rearrangement of molecules. It becomes kinetic when chemical action, *i.e.* rearrangement, takes place.

81. What kind of energy is the energy of compressed air? — *Ans.* Kinetic, since it is due to the motion of the air particles.

82. A body in space is entirely free to move (*i.e.*, free from the influence of all other bodies); how much force will be required to move it? — *Ans.* A body not constrained by other bodies (*i.e.*, perfectly free to move) is perfectly sensitive to the action of a force, so that the smallest force would move the largest mass.

83. Compare the velocities produced on masses of 1^k, 200^g, and 1^g of forces measuring 200,000, 40,000, and 200 dynes. — *Ans.* All equal if applied for the same length of time; 200^{cm} per second if the action endure one second.

84. Given a body in motion. At a given instant let it be left to itself and not acted on by any force. What will happen? (See Maxwell's "Matter and Motion," pp. 56, 57.)

85. (*a*) Which has the greater energy, a body moving at the rate of 20^m per second in a straight line, or one of the same mass moving with the same velocity in a circular path? (*b*) If the force which compels the latter to move in a circular path should cease to act, what would be its subsequent velocity? — *Ans.* (*a*) Their energies would be the same, for energy does not depend on direction or form of path, but on the velocity at each instant along the path. (*b*) The velocity would be 20^m per second.

86. Let two equal forces act for the same length of time, one on a body weighing 2^k, the other on a body weighing 6^k; how will the momenta produced compare?

87. Is a spring balance a force measurer or an energy measurer? Why will it not answer both purposes?

88. How can a power of 5 lbs. raise a ton 10 ft. with a perfect machine?

TEST QUESTIONS.

89. What power will raise 20 tons of coal 100 ft. in an hour?

90. How many times as much energy has a body moving 100 ft. per second than another body of the same weight moving 25 ft. per second? Compare their momenta.

91. How much faster will an iron ball weighing a pound fall than one weighing an ounce?

92. Imagine that a body having a mass of 40^g is at absolute rest in space, and is absolutely free from the action of all external forces. Now let a force of 20 dynes act upon it for five seconds in one direction. What will be the result? Is work done upon the body? (Inertia is not a resistance — is not a force.) When a body offers no resistance to the action of a force, what is the only effect produced by the force? What kind of motion is the result? What kind of energy will the body acquire, i.e., kinetic or potential? What velocity will the body acquire? — *Ans.* $2\frac{1}{2}^{cm}$ per second. What amount of energy will be imparted to the body? — *Ans.* 50 ergs.

93. Is a pendulum which vibrates seconds at New York longer or shorter than one which vibrates seconds at the equator? Explain.

94. Which will tick oftener, a clock having an 8-inch pendulum or one having a 32-inch pendulum? How many times oftener?

95. From the laws of the pendulum derive a reason why a person with short legs naturally takes quicker steps than a person with longer legs.

96. Describe the transformations of energy that take place during a single swing of a pendulum.

97. Where will a given pendulum vibrate faster, at the top or at the foot of a mountain? Why?

98. Which will vibrate in a shorter time, a pendulum 10 inches long or one 15 inches long? How many times shorter?

99. Two clocks — one at the equator, the other at a pole — have pendulums of the same length. Which will gain on the other, and why?

100. It is sometimes necessary to use a pendulum less than a meter in length to beat seconds. How may this be accomplished? — *Ans.* By placing a bob on the pendulum rod above as well as below the center of motion. By moving this bob up and down the rod the pendulum may be made to move slow or fast as is desirable. The musician's *metronome* is an example.

101. A body starts from rest under the influence of a force which produces acceleration $a = 2$ feet; when will it have a velocity of 1000 ft. per second? — *Ans.* At the end of the 500th second.

102. A body travels at 12 ft. per second. In 10 seconds it is moving 7 ft. per second. What is the mean retardation? — *Ans.* $\frac{1}{2}$ foot per second.

103. Wherein is the absolute unit of force preferable to the gravitation unit of force? — *Ans.* The former is not affected by the variations in the force of gravity, and hence is everywhere the same, while the latter is subject to local variation.

104. The final velocity of a falling body weighing 5 lbs. on striking the ground is 100 ft. per second. "With what force will it strike the ground?" — *Ans.* The question as it stands is devoid of sense, for the time during which it acts (depending upon the rigidity of both the body and the earth) is not given. The question may be stated thus: What is the mean pressure between the body which has fallen and the earth on which it falls, if a velocity of 100 ft. per second is arrested in t units of time? Assume that t is $\frac{1}{2000}$ of a second. Since $v = at$; $v = 100$; $t = \frac{1}{2000}$; $a = 200,000$; and $F = ma = \frac{5}{32.2} \times 200,000 = 31,052.7 +$ lbs.

105. What is the velocity of a falling body at the end of the 5th second at a place where $g = 2^m$ per second? — *Ans.* 10^m per second.

106. A bullet is fired from a gun whose barrel is 30 inches long; describe its motion through the first 30 inches. — *Ans.* Its motion is accelerated, because it is acted upon by a continuous (not constant) force throughout this distance.

107. When a body is thrown horizontally into the air, why does it fall to the earth? What effect upon the rapidity and time of falling has its horizontal motion?

108. Suppose that a cubic centimeter of water at 4° C. to become frozen: (a) What will it weigh? (b) What will be its mass? Suppose it to be suspended by a thread: (c) What tenion in the thread will it produce, measured by the gravitation system? (d) What, measured by the absolute system? — *Ans.* (a) Weighed by a balance-beam it will weigh one gram; weighed by a spring balance it will depend upon the locality; (b) its mass is one gram; (c) the same tension that would be produced under the same circumstances if a standard gram-mass (usually of platinum) were suspended, and both would depend upon the locality; (d) measured by the absolute system it would depend upon the locality: at sea level, in the latitude of Greenwich, it would be 981 dynes.

109. Suppose that you take a cubic centimeter of water at 4° C. and allow it to freeze: (a) How will its mass be affected? (b) What is its density before there is a change of temperature? (c) What, after it is frozen? (d) How is its volume affected by the change? — *Ans.* (a) Its mass will not be changed; (b) its density is 1; (c) its density will be less than 1; (d) its volume will be increased.

110. In the metric system what is the density of a body? — *Ans.* It is the number of grams in a cubic centimeter.

111. Given a solid, a vessel of water and a vessel of another liquid, and a pair of balances; state how you would find the specific gravity of the solid, the liquid, and the cubical contents of the solid.

112. State three methods of finding the specific gravity of a liquid.

113. Suggest an easy method of finding the cubical contents of a test-tube. — *Ans.* Ascertain the weight of the water it contains, and the weight in grams equals its contents in cubic centimeters.

114. A pebble-stone weighs in air 20^g; immersed in water it weighs 15^g; immersed in another liquid it weighs 17^g. What is the specific gravity of the latter liquid? What is the specific gravity of the stone? What is the cubical contents of the stone?

115. What is heat? Give some proof of your statement.

116. How will you explain the rush of air into the vacuum when an opening is made into an exhausted air-pump receiver?

117. What is the difference between a hot body and a cold body?

118. Why is the quantity of heat required to raise the temperature of a body of gas 1° very different when it is in an open vessel to what it is in a closed one? — *Ans.* When gases are heated in an open vessel, they expand very rapidly, and a considerable portion of the heat is expended in changing their bulk; in a closed vessel, the whole of the heat is spent in raising the temperature.

119. Name several processes by which the temperature of a body may be lowered without removing heat from it? — *Ans.* Expansion, evaporation, and liquefaction.

120. For what purpose is ice wrapped in flannels in the summer? — *Ans.* To exclude the heat.

121. Let a kilogram of mercury lose one calorie; how much will its temperature be lowered?

122. How high must a body be raised that on falling it will generate enough heat to raise its own weight of water 1° C.? — *Ans.* Suppose the body weighs one pound; then it must be raised $772 \times \frac{9}{5}$ = about 1390 ft. (See § 147, Physics.)

123. How many kilograms of ice at 0° C. can be melted by 1^k of steam at 100° C.? — *Ans.* $(537 + 100) \div 80 = 7.9^k+$.

124. How many kilograms of steam at 100° C. will melt 100^k of ice at 0° C.? — *Ans.* $(100 \times 80) \div (537 + 100) = 12.5^k+$.

125. What weight of steam at 100° C. would be required to raise 500^k of water from 0° C. to 10° C.?

Ans. $(500 \times 10) \div (537 + 90) = 7.9^k+$.

126. A current of 9 ampères worked on an electric arc-light, and, on measuring the difference of potential between the two carbons by an electrometer, it was found to be 140 volts. What was the amount of horse-power absorbed by this lamp?

Ans. $\dfrac{C \times V}{745} = \dfrac{9 \times 140}{745} = 1.66$ horse-powers. (See rule X., p. 69.)

127. How many incandescent lamps, requiring an E.M.F. of 60 volts and a current of 1.5 ampères each, can be supplied by an engine giving 15 useful horse-powers, the loss of energy in the dynamo being 20 per cent?—*Ans.* 80 per cent of $15 = 12$, the horse-power of current available. From the tables, p. 76, 1 horse-power = (about) 746 volt-ampères, 1 lamp requires $1.5 \times 60 = 90$ volt-ampères; therefore, $\dfrac{12 \times 746}{90} = 100$ lamps, very nearly.

128. What amount of heat will be generated in each of the foregoing lamps per second?

Ans. $0.00024 \times (1.5 \times 60) = 0.00216$ calorie of heat.

129. An electric bell is in circuit with a voltaic cell which will furnish a current just sufficient to ring it. What will be necessary if ten such bells are introduced into the circuit? Why?—*Ans.* If a given current will ring one bell, it will ring any number of like bells. But as each bell introduced into the circuit increases the total resistance of the circuit, the E.M.F. must be correspondingly increased by the introduction of new cells in series, in order to maintain the same current.

130. Why does it require more voltaic cells to work a long telegraph line than a short one? How ought they to be connected? Why?

131. For each 50 ohms' resistance in a circuit, about 1 gravity-cell is required (Haskins). Suppose a line of wire 200 miles long (13 ohms' resistance to the mile), and 10 relays in the circuit: (*a*) What should be the theoretical resistance of each relay? (*b*) How many gravity-cells will be required to operate them?—*Ans.* (*a*) Disregarding the resistance of the battery (its resist-

174   TEST QUESTIONS.

ance in this case being relatively of no importance), the resistance of the circuit is $13 \times 200 = 2600$ ohms; hence, the resistance of the relays should be 2600 ohms, or 260 ohms in each relay. (b) The entire external resistance is $2600 + 2600 = 5200$ ohms; $5200 \div 50 = 104$, the number of cells required.

132. Ten Bunsen cells, whose E.M.F. $= 1.7$ and $r = 0.5$ ohm, are to be used in a circuit whose $R = 2$ ohms: (a) What will be the current if they are connected abreast? (b) What, if they are connected tandem? (c) What, if they are joined in pairs abreast, and the pairs are connected with one another tandem? (d) What, if they are divided into two groups of five each, the cells of each group connected abreast, and the two groups are connected tandem?

Ans. (a) $\qquad C = \dfrac{E}{r+R} = \dfrac{1.7}{0.05+2} = 0.82$ ampère.

(b) $\qquad \dfrac{17}{5+2} = 2.41+$ ampères.

(c) $\qquad \dfrac{8.5}{0.25 \times 5 + 2} = 2.6+$ ampères.

(d) $\qquad \dfrac{3.4}{0.1 \times 2 + 2} = 1.72+$ ampères.

133. The resistance of 5 inches of No. 32 platinum wire is about 0.05 ohm; how would you connect 4 Bunsen cells so as to develop in the wire the maximum quantity of heat?

Ans. $\sqrt{nr \div R} = \sqrt{4 \times 0.5 \div 0.05} = 6+$. The interpretation of this is, that all should be connected abreast.

134. What is the maximum amount of heat that can be developed in the wire per second with the four cells?

Ans. $C = \dfrac{E}{r+R} = \dfrac{1.8}{0.125+0.05} = 10.2+$ amperes.

$H = C^2 \times R \times t \times 0.00024 = 10.2^2 \times 0.05 \times 1 \times 0.00024$
$\qquad\qquad\qquad\qquad = 0.001248$ calorie;

or, sufficient heat can be developed to raise the temperature of 1.248^{cc} of water 1° C. per second.

TEST QUESTIONS. 175

135. What current will there be when 10 gravity-cells (E.M.F. = 1 volt each, and $r = 3$ ohms each) are connected in series through a wire whose resistance is 50 ohms?

$$\text{Ans. } C = \frac{E}{r + R} = \frac{10}{30 + 50} = 1.25 \text{ ampère.}$$

136. Show in the preceding question that, with an infinite number of cells in series, the current cannot possibly exceed $3\frac{1}{3}$ ampères.

Ans. Since the external resistance in this case will become of comparatively no importance, it may be disregarded; then

$$C = \frac{E}{r} = \frac{1 \times \text{infinity}}{3 \times \text{infinity}} = 0.3\frac{1}{3} \text{ ampère.}$$

137. It is required to ring a bell over a No. 16 copper wire 300 ft. long, with three cells of Leclanché battery, the resistance of the wire being 0.0076 ohm per yard, and the resistance of a Leclanché cell being 1 ohm. What should be the resistance of the bell magnet to obtain the greatest magnetic power? — Ans. 0.76 ohm (the resistance of 300 ft. of wire) + 3 ohms (the resistance of three cells) = 3.76 ohms. The resistance of the circuit, not including the helix of the electro-magnet, is, therefore, 3.76 ohms; hence (Law 1 of electro-magnets), the resistance of the bell magnet should be 3.76 ohms.

138. What is the resistance of the carbon filament of an incandescent light in which there is a fall of potential of 60 volts, and the intensity of the current is 1.6 ampères?

$$\text{Ans. } R = \frac{E}{C} = \frac{60}{1.6} = 37\frac{1}{2} \text{ ohms.}$$

139. Explain the sparks seen at the circuit-breaker of an induction coil when in operation. — Ans. They are sparks produced by the extra currents at each " breaking " of the circuit.

140. A line 2 miles long, built of No. 8 iron wire whose resistance is 13 ohms per mile, has two bell magnets in circuit, and a battery of 10 Leclanché cells (r of each = 1 ohm). What should be the theoretical resistance of each bell magnet?

—*Ans.* Battery resistance 10 ohms + line resistance 26 ohms = 36 ohms. The sum of the resistances of the electro-magnets should then be 36 ohms, or 18 ohms each. But, as there would be likely to be some leakage, practically the resistance of each magnet should be some less than 18 ohms.

141. For which is the gravity battery better adapted, circuits of small or large resistance? Why?—*Ans.* For circuits of large resistance, since the large resistance of the battery then becomes of comparatively little importance.

142. You have 48 cells, each of 1.2 volt E.M.F., and each of 2 ohms' internal resistance. What is the best way of grouping them together when it is desired to send the strongest possible current through a circuit whose resistance is 12 ohms?— *Ans.* Group them three abreast. (See Law VIII., p. 69.)

143. Required the current in a circuit with 60 Grove cells, connected in series with 12 ohms' external resistance; $r = 0.6$ ohm, and $E = 1.8$ volts for each cell.

Ans. $C = \dfrac{108}{36 + 12} = 2.25$ ampères.

144. Required the current in the same circuit when the arrangement of the 60 cells is 30 series of 2 cells joined abreast.

Ans. $C = \dfrac{54}{9 + 12} = 2.56+$ ampères.

145. Required the current in a circuit, with the same 60 cells connected in series, when $R = 2$ ohms.

Ans. $C = \dfrac{108}{36 + 2} = 2.85+$ ampères.

146. Required the current in the last circuit with the arrangement of 12 cells in series, each consisting of 5 cells connected abreast. *Ans.* $C = \dfrac{21.6}{1.44 + 2} = 6.27$ ampères.

147. How many Bunsen cells (E.M.F., 1.7 volts per cell) will be required to maintain an electric arc-light, whose resistance is 8 ohms, with a current of 9 ampères?

TEST QUESTIONS. 177

Ans. $C = \dfrac{E}{R}$; or, $E = CR = 9 \times 8 = 72$ volts; $72 \div 1.7 = 42 +$; hence, about 42 cells will be required.

148. What is the cause of a current of electricity?—*Ans.* A difference of potential at different points in the conductor through which it flows.

149. If two Bunsen cells are to be used in a circuit, with an external resistance of 2 ohms, should they be connected abreast or tandem? If two gravity cells should be used in the same circuit, by which method should they be connected?

150. What E.M.F. is required to maintain a current of 20 ampères in a circuit of 100 ohms' resistance?

Ans. $C = \dfrac{E}{R}$; whence $E = C \times R = 20 \times 100 = 2000$ volts.

151. What current will a battery having an E.M.F. of 4 volts furnish in a circuit whose total resistance is 10 ohms?

Ans. $C = \dfrac{E}{R} = \dfrac{4}{10} = 0.4$ ampère.

152. What is the total resistance of a circuit in which a battery having an E.M.F. of 2 volts furnishes a current of 0.5 ampère?

Ans. $C = \dfrac{E}{R}$; whence $R = \dfrac{E}{C} = \dfrac{2}{0.5} = 4$ ohms.

153. What is the most convenient test of the E.M.F. of an electrical machine?—*Ans.* The length of the sparks which it will give.

154. If a sounding body moves, how will its motion affect the wave-length of the waves which it throws behind? How will it affect those thrown in front? How will the pitch of the sound compare as heard by a person in front and another behind?

155. When a voice an octave higher, such as that of a woman or boy, reproduces the same melody which has been sung by a man, we "hear again a part of what we have heard before." Explain.—*Ans.* A human voice conveys to the hearer not only the primes of the compound tones, but also their upper octaves,

and with less force the other upper overtones; hence a voice an octave higher would produce the upper octave previously given by the man, or a "part" of what was previously given.

156. What kind of vibration is that of a column of air in a pipe?—*Ans.* Longitudinal.

157. The picture on a stereopticon slide is two inches square. The slide is ten inches from the lens of a porte lumière. What will be the size of the image on the screen at a distance of 30 ft.?—*Ans.* $\frac{O}{I} = \frac{x}{2}$, or $\frac{360}{10} = \frac{x}{2}$; whence $x = 72$ in. $= 6$ ft., *i.e.*, the image is 6 ft. square.

158. What is the focal length of the lens used in the last question?

Ans. $\frac{1}{o} + \frac{1}{i} = \frac{1}{f}$, or $\frac{1}{360} + \frac{1}{10} = \frac{1}{f}$; whence $f = 9.7 +$ in.

159. At what temperature does a body cease to radiate heat and light?—*Ans.* It ceases to radiate heat at the absolute zero; light at about 525° C.

160. What phenomenon shows that light does, in a small degree, pass around a corner?—*Ans.* Diffraction.

161. Why does a white body always appear of the same color as the light by which it is illuminated?

162. What proof can you give that the light of the electric spark does not proceed from an incandescent electric fluid (if there be such a substance)' nor any etherial medium which is supposed to pervade all space?—*Ans.* Every line found in the spectrum of the light proceeding from the electric spark can be traced to some chemical substance existing either in the electrodes or in the space through which the electricity passes, and there are none that are common to all discharges, as would be the case if a common medium were rendered luminous.

163. A gas-burner must have what candle-power in order that it may illuminate a printed page as brightly at a distance of 5 ft. as a single candle at a distance of 1 ft.?—*Ans.* 25 candle-power.

164. How does one color differ from another color?

165. How do you explain the separation of colors when white light passes through an optical prism?

166. What is the general effect of a concave mirror on a beam of light? Name some other piece of optical apparatus that will produce the same effect.

167. Why is the image of a light as seen in water usually enormously elongated vertically?

168. Why is the same side of the moon always turned toward the earth? Does the moon rotate on its axis?

Part V.

SOLUTIONS TO PROBLEMS IN ELEMENTS OF PHYSICS.

CHAPTER II.

DYNAMICS.

Page 52. Q. 4. $76 : 49.2 :: 1033.3 : 668.92^g +$.

Q. 5. $76 : 98.2 :: 1033.3 : 1335.13^g +$.

Page 60. Q. 8. Any weight less than 10^k may be lifted.

Page 68. Q. 3. The pressure on the top is nothing; on the bottom it is $25 \times 20 \times 15^g = 7500^g$; on each of the sides it is $25 \times 15 \times 7.5^g = 2812.5^g$; and on the ends, $20 \times 15 \times 7.5^g = 2250^g$.

Q. 4. The additional pressure will be 100^g for every 4^{qcm} of area on the inner surface. The area of the bottom is 500^{qcm}; the additional pressure is, therefore,

$$\frac{500}{4} \times 100^g = 12,500^g.$$

This, also, is evidently the pressure on the top. The additional pressure on each of the sides is

$$\frac{375}{4} \times 100^g = 9375^g;$$

and on each of the ends,

$$\frac{300}{4} \times 100^g = 7500^g.$$

PAGE 69. Q. 5. The total pressure on the bottom is 7500^g $+ 12,500^g = 20,000^g$; that on the top, $12,500^g$; on each side, $12,187.5^g$; on each end, 9.750^g.

Q. 6. The answers to Q. 3 would be 13.6 times greater: viz., $102,000^g$ on the bottom; $38,250^g$ on each side, and $30,600^g$ on each end.

In considering Q. 4, it makes no difference whether mercury or water is used. In the case of mercury, the total pressure on the bottom is evidently $102,000^g + 12.500^g = 114,500^g$; on the top. $12,500^g$; on each side, $38,250^g + 9375^g = 47,625^g$; and on each end, $30.600^g + 7500^g = 38,100^g$.

Q. 7. (*a*) The pressure on the bottom of the keg, when the tube is empty, is evidently 1200^g. (*b*) If the tube be filled, the column of water will be 1030^{cm} high, instead of 30^{cm}; therefore the pressure on the bottom is $40 \times 1030 = 41,200^g$. (*c*) The weight of the water in the tube is evidently 1000^g.

Q. 8. Assuming, for simplicity, that all parts of the box are at equal depths, the crushing force on each side would be equal to the weight of a column of water 1^{km} high, with a base of 1^{qm}; the volume of this is 1000^{cbm}, and its weight $1,000,000^k$.

Q. 9. Taking the atmospheric pressure at 1^k per square centimeter, the crushing force at the sea level, on each side, would be $10,000^k$.

Q. 10. Since the whole area of the top is 500^{qcm}, a pressure of 20^g on the plug would make a total pressure of

$$\frac{500}{4} \times 20^g = 2500^g :$$

but, by the conditions of the problem, the top can sustain 50^g on each 10^{qcm}, or 2500^g total; therefore, it is clear that any pressure on the plug greater than 20^g would burst the vessel.

PAGE 84. Q. 1. $1033.3 \div 1.841 = 561.27^{cm} +$.

Q. 2. The 50^g of water when immersed in water is evidently buoyed up with a force of 50^g, and no weight is indicated.

CHAPTER II. — DYNAMICS. 183

Q. 3. The combined solids displace 102.88^{cc}. The sinker alone displaces 14^{cc}; $102.88^{cc} - 14^{cc} = 88.88^{cc}$; hence,

$$G = \frac{W}{W_1} = \frac{80}{88.88} = 0.9 +.$$

Q. 4. The weight of the water displaced, or the buoyant force when the oil is completely immersed, is greater than the weight of the oil; we have, then, two unequal forces in opposite directions, and the oil rises until the weight of the water displaced just equals the weight of the oil.

Q. 5. In the case of a floating tumbler, it will be noticed that by far the larger part of the water displaced is displaced not by the glass simply, but by the air in the bottom of the tumbler, and the *average density* of the combination of air and glass is less than the density of water, so the tumbler floats.

Q. 6. Iron vessels float for the same reason that the tumbler does.

Q. 7. From the table of specific gravities, we find that 1^{cc} of ice weighs 0.92^{g}, then 500^{cc} weigh 460^{g}; 460^{g} or 460^{cc} of water will be displaced, so that $500^{cc} - 460^{cc}$, or 40^{cc}, of ice will be above the surface.

Q. 8. Ice will sink in alcohol, since its specific gravity is greater than that of alcohol.

Q. 9. The weight of 500^{cc} of fresh water is 500^{g}, that of 500^{cc} of sea water is $500 \times 1.026^{g} = 513^{g}$, making 13^{g} more matter in the sea water than in the fresh water.

Q. 10. $\frac{50000}{19.36} = 2582.64^{cc} +.$

Q. 11. 19.36^{g} per cubic centimeter.

Q. 12. 0.24^{g} per cubic centimeter.

Q. 13. $\frac{1}{773}^{g}$ (0.0012932^{g}) per cubic centimeter.

Q. 14. The 53^{g} lost weight is the weight of an equal bulk of water, so the volume of the marble is 53^{cc}.

Q. 16. $\dfrac{1}{0.0012932} = 773.27^{cc} +$.

PAGE 85. Q. 17. $G = \dfrac{W}{W'}$, or $8.79 = \dfrac{1000}{W'}$;

∴ $W' = \dfrac{1000}{8.79} = 113.76^g =$ weight of an equal bulk of water. The piece of copper, therefore, weighs in water $1000^g - 113.76^g = 886.24^g$.

Q. 18. The cubical contents is $20 \times 10 \times 5^{cm} = 1000^{cc}$; therefore the weight is $1000 \times 11.35 = 11{,}350^g$.

Q. 19. It will lose the weight of an equal bulk of water, viz., 1000^g; thus weighing, when immersed, $10{,}350^g$.

Q. 20. Lead will float on the surface of mercury.

Q. 21. The weight that is lost is transferred to the liquid. (See Exp. 2, p. 76.)

Q. 22. $\dfrac{1015}{1000} = 1.015$.

Q. 23. $V = \dfrac{W}{D} = \dfrac{1000}{11.35} = 88.10^{cc} +$; the weight of an equal volume of air is $88.1 \times 0.0012932^g = 0.11393092^g + =$ the weight gained by weighing in a vacuum; therefore, the weight in a vacuum is $1000.11393092^g +$.

Q. 24. The specific gravity of the other liquid is
$$\dfrac{30 - 27}{30 - 26} = \dfrac{3}{4} = 0.75.$$

Q. 25. $G = \dfrac{W}{W'}$, or $W' = \dfrac{W}{G} = \dfrac{150}{10.47} = 14.32^g + =$ the weight supported by the water; $150^g - 14.32^g = 135.68^g +$, the weight supported by the string.

Q. 26. The weight of the boat is evidently the weight of 25^{cbm} of water, or $25{,}000^k$.

Q. 27. It would displace 50^k of water more, viz., $25{,}050^k$.

CHAPTER II. — DYNAMICS. 185

Q. 28. 100^{cbm} of water weighs $100,000^k$; the boat alone weighs $25,000^k$; therefore, it will take $75,000^k$ to sink the rail to the water-level.

Q. 29. $\dfrac{105.928 - 100}{102.4 - 100} = \dfrac{5.928}{2.4} = 2.47.$

Q. 30. 1^l of water weighs 1000^g; the density of alcohol is 0.8; therefore, 1^l of alcohol weighs 800^g.

Q. 31. $V = \dfrac{W}{D} = \dfrac{50}{1.841} = 27.15^{cc}+.$

Q. 32. $V = \dfrac{W}{D} = \dfrac{80}{1.42} = 56.33^{cc}+.$

Q. 33. $W = V \times D = 35 \times 0.847 = 29.645^g.$

Q. 34. Each square centimeter must be able to sustain the weight of a column of water 2000^{cm} high, or $2000^g = 2^k$ per square centimeter.

Q. 35. The bottom sustains $2500 \times 50 = 125,000^g = 125^k$; each side sustains one-half of this, viz., 62.5^k.

Q. 36. It will sink a little way, for the buoyant effect of the air on the part not immersed in the liquid will be removed.

PAGE 96. Q. 1. Let x represent the distance from the boy that the weight should be placed,

$x : 3 - x :: 30 : 20$; whence, $x = 1.8^m$.

Q. 2. $40 : 260 :: 50 - x : x$; or, $x = 43.33^k+ =$ weight supported by the man; so the boy's load is 6.66^k+.

Q. 3. Half a mile down the stream.

Q. 4. $\frac{1}{2}\sqrt{2}$ miles.

Q. 5. Half an hour.

Q. 6. $\frac{1}{2}\sqrt{2} \times 10 = 5\sqrt{2}$ miles per hour.

PAGE 107. Q. 1. $S = \frac{1}{2}gT^2 = \frac{1}{2} \times 9.8^m \times 25 = 122.5^m$
$= \frac{1}{2} \times 32\frac{1}{6}$ ft. $\times 25 = 402.08 +$ ft.

Q. 2. $s = \frac{1}{2}g(2T-1) = \frac{1}{2} \times 9.8^m \times 9 = 44.1^m$
$= \frac{1}{2} \times 32\frac{1}{6}$ ft. $\times 9 = 144.74 +$ ft.

Q. 3. $V = gT = 9.8^m \times 5 = 49.0^m = 32\frac{1}{6}$ ft. $\times 5 = 160.83 +$ ft.

Q. 4. $S = \frac{1}{2}gT^2 = \frac{1}{2} \times 9.8^m \times 49 = 240.1^m = \frac{1}{2} \times 32\frac{1}{6}$ ft. $\times 49$
$= 788.08 +$ ft.

Q. 5. $S = \frac{1}{2}kT^2 = 500^m \times 3600 = 1,800,000^m$
$= 1640.42$ ft.* $\times 3600 = 5,905,512 +$ ft.

Q. 6. $V = \frac{1}{2}k \times 2T = 500^m \times 60 = 30,000^m$ per minute
$= 1640.42$ ft. $\times 60 = 98,425.2$ ft. per minute.

Q. 7. $s = \frac{1}{2}k(2T-1) = 500^m \times 117 = 58,500^m$
$= 1640.42$ ft. $\times 117 = 191,929.14 +$ ft.

Q. 8. $S = \frac{1}{2}kT^2 = 2^m \times 16 = 32^m = 6.56 +$ ft. $\times 16 = 104.96$ ft. from a point directly under that from which it started.

Q. 9. $V = \frac{1}{2}k \times 2T = 2^m \times 8 = 16^m = 6.56$ ft. $\times 8 = 52.48$ ft. per second.

Q. 10. $V = gT = 9.8^m \times 4 = 39.2^m = 32\frac{1}{6} \times 4 = 128.66 +$ ft. per second.

Q. 11. $V = gT = 9.8^m \times 3 = 29.4^m = 32\frac{1}{6} \times 3 = 96.5$ ft. per second.

Q. 12. It will rise in the first second as far as it would fall in the third.
$s = \frac{1}{2}g(2T-1) = \frac{1}{2} \times 9.8^m \times 5 = 24.5^m = \frac{1}{2} \times 32\frac{1}{6}$ ft. $\times 5$
$= 80.41 +$ ft.

PAGE 111. Q. 1. They would vibrate in equal time, since the accelerative effect of gravity on all bodies is the same at the same place.

Q. 2. $1 : \frac{1}{2} :: \sqrt{0.993} : \sqrt{x}$, or $x = 0.248^m$
$=$ the length of a pendulum beating half-seconds.

$1 : \frac{1}{4} :: \sqrt{0.993} : \sqrt{x}$, or $x = 0.062^m$
$=$ length of one beating quarter-seconds.

$1 : 2 :: \sqrt{0.993} : \sqrt{x}$, or $x = 3.972^m$
$=$ length of one beating once in two seconds.

* 1 foot $= 0.3048^m$.

CHAPTER II. — DYNAMICS.

$1 : 30 :: \sqrt{0.993} : \sqrt{x}$, or $x = 893.7^m$
= length of one beating once in thirty seconds or twice each minute.

Q. 3. $1 : \dfrac{60}{x} :: \sqrt{0.993} : \sqrt{0.4}$ where both lengths are expressed in meters, and x is the required number.

PAGE 116. Q. 1. Momentum = mass × velocity = 100,000 × ⅙ = 16,666.66 + for the car, where the mass is expressed in pounds and the velocity in feet per second. The momentum of the ice = 500 × 96.5 = 48,250; so the momentum of the ice is nearly three times that of the car.

Q. 3. $25x = 80 \times 10 = 800$; therefore $x = 32^{km}$ per hour.

Q. 4. To double the momentum with a constant mass, the velocity must be doubled; to double the velocity the time must be doubled; but, by doubling the time a body is falling, the space is increased four-fold.

PAGE 130. Q. 5. The work $= 80 \times 4 \times 60 = 19,200^{kgm}$ per hour.

Q. 6. (a) Falling freely for 4 seconds, a body would fall through a space $S = \tfrac{1}{2}gT^2$, or $4.9 \times 16 = 78.4^m$; therefore, in order that a body weighing 50^g may rise 78.4^m, energy equal to 0.05×78.4, or 3.92^{kgm}, must be imparted to it.

(b) Here $S = 4.9 \times 25 = 122.5^m$, and the energy will be 0.05×122.5 or $6.125^{kgm} = \tfrac{25}{16}$ of 3.92^{kgm}; *i.e.*, the energy required to cause a body to rise 5 seconds is $\tfrac{25}{16}$ of that required to cause it to rise 4 seconds.

(c) The reason of this is that the initial velocity necessary to enable a body to rise 5 seconds is $\tfrac{5}{4}$ of that which would enable it to rise 4 seconds, and the energy increases as the square of the velocity.

Q. 7. Since the momentum of any moving body is proportional to the velocity; and since, in falling bodies, the velocity varies as the time, it follows that the momentum in the case (b) above is $\tfrac{5}{4}$ of that in (a).

188 SOLUTIONS TO PROBLEMS.

Q. 8. Since the energy = weight into hight (Ws), the energy stored in 50^k 80^m high is $50 \times 80 = 4000^{kgm}$.

Q. 9. Energy $= \dfrac{WV^2}{2g} = \dfrac{50 \times 100^2}{2 \times 9.8} = 25{,}510.20^{kgm} +$.

Q. 10. $S = \tfrac{1}{2}gT^2 = 4.9 \times 16 = 78.4^m$;
energy $= WS = 50 \times 78.4 = 3920^{kgm}$.

Q. 11. If the 50^k should fall in air, a part of its energy would be transformed into heat.

Q. 12. Energy $= \dfrac{WV^2}{2g} = \dfrac{25 \times (29.4)^2}{2 \times 9.8} = 1102.5^{kgm}$.

Q. 13. During the ascent its energy is expended in doing work by lifting the 25^k to a hight against the force of gravity.

Q. 14. (a) Momentum $= MV = 50 \times 2 = 100$; again, for the 50^g, with a velocity of 100^m per second, we have : momentum $= MV = 0.05 \times 100 = 5$; therefore 50^k moving 2^m per second has 20 times the momentum of 50^g moving 100^m per second.

(b) Energy $= \dfrac{MV^2}{2g} = \dfrac{50 \times (2)^2}{2 \times 9.8} = 10.204^{kgm}$;

again, energy $= \dfrac{MV^2}{2g} = \dfrac{0.05 \times (100)^2}{2 \times 9.8} = 25.51^{kgm}$;

i.e., the energy in the second case is $\tfrac{5}{2}$ of that in the first.

Q. 15. Energy is the power of doing work; work is the overcoming of resistance through space; therefore it is its energy that enables one to determine the amount of resistance that a moving body can overcome.

Q. 16. A child can draw a carriage weighing 150^k because the energy required to overcome the resistance offered by the revolving wheels is much less than that required to raise 30^k against the force of gravity.

Q. 17. (a) The work of the horse is equivalent to that of raising 40^k 100^m per minute $= 4000^{kgm}$ per minute.

(b) Since 1 horse-power $= 4570^{kgm}$ per minute, 4000^{kgm} per minute $= \dfrac{4000}{4570}$ h.p. $= 0.87 +$ h.p.

CHAPTER II. — DYNAMICS.   189

Q. 18. 3^{km} per hour $= 50^m$ per minute; the power required to move 70^k 50^m per minute is 3500^{kgm} per minute $= \dfrac{3500}{4570}$ h.p. $= 0.76 +$ h.p.

Q. 19. 5^m per hour $= \frac{1}{12}^m$ per minute; to raise $1,350,000^k$ $\frac{1}{12}^m$ will require $\frac{1}{12}$ of $1,350,000^{kgm}$ of work per minute $= 112,500^{kgm}$ $= \dfrac{112500}{4570}$ h.p. $= 24.61 +$ h.p.

Q. 20. 10 tons $= 20,000$ lbs.; a 3 h.p. engine will raise 99,000 lbs. 1 ft. in 1 minute, or $\frac{1}{50}$ of $99,000 = 1980$ lbs. 50 ft. in 1 minute; therefore, to raise 20,000 lbs. 50 ft., it will take $\dfrac{20000}{1980} = 10.10 +$ minutes.

Q. 21. A 2 h.p. engine will raise $2 \times 4570^k = 9140^k$ 1^m per minute; in 10 seconds the same engine will raise $\frac{1}{6} \times 9140^k = 1523\frac{1}{3}^k$ 1^m; therefore, in the same time it will raise 1000^k $\dfrac{1523\frac{1}{3}^m}{1000} = 1.523^m +$.

Q. 22. A 5 h.p. engine can do 5×4570^{kgm} of work per minute, or $60 \times 5 \times 4570^{kgm} = 1,371,000^{kgm}$ per hour.

Q. 23. $\dfrac{1371000}{90000} = 15.23 +$ days.

Q. 24. The energy increases as the square of the velocity; therefore, to increase the energy four-fold, the velocity must be doubled.

PAGE 135. Q. 1. (*a*) Since the power, multiplied by the distance through which it moves, must equal the weight multiplied by its distance, it is clear that the 10^k will be moved with a velocity of 2^m per second.

(*b*) $\frac{20}{50} = 0.4^m$ per second.

Q. 2. $Pp = Ww$; ∴ $50 \times 100 = 2W$, or $W = 2500^k$. The advantage would be one of convenience, since a small power moving through a considerable distance can move a very great weight through a short distance. In common parlance we should say that "power" is gained.

Q. 6. $\dfrac{W}{P} = \dfrac{p}{w}$, or $\dfrac{4}{2} = \dfrac{p}{75-p}$; $\therefore p = 50^{cm}$, *i.e.*, the prop must be 50^{cm} from the power end. The pressure on the prop will be, clearly, $2 + 4 = 6^k$ added to the weight of the lever.

Q. 9. Suppose the prop to be x^{cm} from the end from which 5^k are suspended, then $5x = 20(70-x)$, or $x = 56^{cm}$.

Q. 11. $3W = 15 \times 1$, or $W = 5$ lbs.

PAGE 136. Q. 12. $6 \times 3 = 1 \times x$, or the number of spaces of P from the fulcrum is 18.

Q. 13. The power multiplied by its distance must equal the weight multiplied by its distance, *i.e.*, $240 P = 60 \times 40$, or $P = 10$.

Q. 14. $10^m = 1000^{cm}$; since the circumference of the axle $= 60^{cm}$, it will take $\dfrac{1000}{60} = 16\tfrac{2}{3}$ turns to raise the bucket from the cavity. The power at each turn travels 240^{cm}; therefore, the whole distance that the power must travel is $16\tfrac{2}{3} \times 240^{cm} = 4000^{cm} = 40^m$.

Q. 15. (a) $\dfrac{W}{P} = \dfrac{p}{w}$, or $\dfrac{W}{1} = \dfrac{36}{4}$; $\therefore W = 9$ lbs.

(b) Again, $\dfrac{W'}{9} = \dfrac{30}{6}$; $\therefore W' = 45$ lbs.

(c) Again, $\dfrac{W''}{45} = \dfrac{40}{8}$; $\therefore W'' = 225$ lbs.

(d) $W'' \times$ its velocity $= P \times$ its velocity; $\therefore P$'s velocity would be 225×5 ft. $= 1125$ ft. per second.

Q. 16. The action of the wheel and axle is the same as that of a lever of which the fulcrum is at the centre of the axle;

$$\therefore \dfrac{P}{W} = \dfrac{\text{radius of axle}}{\text{radius of wheel}};$$

if the axle is a pinion, and the wheel has teeth by which it is turned by another wheel, then, since the circumferences are to each other as their radii, $\dfrac{P}{W} = \dfrac{N'}{N}$, where $N =$ the number of teeth on the wheel, and $N' =$ the number on the axle.

CHAPTER III.— HEAT. 191

Q. 17. Applying the same principle that we have used so often, we have $\frac{W}{P} = \frac{p}{w}$, where p and w mean vertical distances. In the case of the inclined plane, $p = L$ and $w = H$; therefore, the general formula becomes $\frac{W}{P} = \frac{L}{H}$.

Q. 18. $\frac{W}{P} = \frac{L}{H}$; $\therefore \frac{200}{P} = \frac{12}{3}$, or $P = 50$lbs.

Q. 19. As in the case of all simple machines, so in the screw $Pp = Ww$, *i.e.*, $25 \times 14\pi = \frac{1}{4}W$; whence, W, or the pressure beneath the screw, is 4398.24 lbs. We have supposed here that the parts move without friction.

PAGE 137. Q. 20. Here P moves through 10^{cm}, and W through $100 - 98 = 2^{cm}$; therefore, a force of 80^g applied in the direction cd will exert a lateral pressure of $\frac{10}{2} \times 80^g = 400^g$.

CHAPTER III.

HEAT.

PAGE 148. Q. 4. The cubical contents of the room is $3 \times 3 \times 2.5^m = 22.5^{cbm} = 22,500^{cdm}$; since each person breathes 9^{cdm} of air per minute, two persons will be supplied $\frac{22500}{18}$ minutes $= 1250$ minutes $= 20\frac{5}{6}$ hours.

Q. 5. For 1000 persons 1000^{cbm} of *fresh* air is needed to keep the whole mass from becoming vitiated; the room contains $(35 \times 18 \times 7^{cbm}) = 4410^{cbm}$; therefore a complete change once in 4.41 minutes is necessary.

PAGE 153. Q. 1. Since $1°$ C. $= \frac{9}{5}$ of $1°$ F., $80°$ C. $= \frac{9}{5}$ of $80°$ F. $= 144°$ F.

Q. 2. Since $1°$ F. $= \frac{5}{9}$ of $1°$ C., $30°$ F. $= \frac{5}{9}$ of $30°$ C. $= 16\frac{2}{3}°$ C.

192   SOLUTIONS TO PROBLEMS.

Q. 3. (a) The temperature of the room, after the fall, was clearly $68° - 30° = 38°$ F.

(b) C. $= \frac{5}{9}$ (F. $- 32) = \frac{5}{9} (68° - 32°) = 20°$ C. before the fall; and $\frac{5}{9} (38° - 32°) = 3\frac{1}{3}°$ C. after.

PAGE 154. Q. 4. $\frac{9}{5}$ C. $+ 32 =$ F. ;

∴ 100° C. = 212° F. $- 20°$ C. $= - 4°$ F.
   40° $= 104°$ $- 40°$ $= - 40°$
   56° $= 132\frac{4}{5}°$ 80° $= 170°$
   60° $= 140°$ 150° $= 302°$
    0° $= 32°$

Q. 5. $\frac{5}{9}$ (F. $- 32) =$ C. ;

∴ 212° F. = 100° C. $- 10°$ F. $= - 23\frac{1}{3}°$ C.
   32° $=$ 0° $- 20°$ $= - 28\frac{8}{9}°$
   90° $= 32\frac{2}{9}°$ $- 40°$ $= - 40°$
   77° $= 25°$ 40° $= 4\frac{4}{9}°$
   20° $= - 6\frac{2}{3}°$ 59° $= 15°$
   10° $= - 12\frac{2}{9}°$ 329° $= 165°$

PAGE 156. Q. 1. Absolute temperature equals
           C. $+ 273° =$ F. $+ 460°$;
therefore, mercury boils at
           $(350° + 273°)$ C. $= 623°$ C.,
or $(662° + 460°)$ F. $= 1122°$ F.

Mercury freezes at $(-38.8° + 273°)$ C. $= 234.2°$ C.,
or $(-37.8° + 460°)$ F. $= 422.2°$ F. $=$ abs. temp.

Q. 2. (a) The increase in volume will be
           $\frac{75}{273}$ of $500 = 137.36^{cc} +$;
therefore, the total volume at 75° C. $= 637.36^{cc} +$.

(b) In this case the decrease will be
           $\frac{20}{273}$ of $500^{cc} = 36.63^{cc} +$,
so that the final volume will be
           $(500 - 36.63^{cc} +) = 463.37^{cc} +$.

PAGE 157. Q. 4. 30° C. $= 303°$ absolute temperature, and $-15°$ C. $= 258°$ absolute temperature ; $1^l = 1000^{cc}$;

∴ $303 : 258 :: 1000 : 851.48^{cc} +$.

CHAPTER III. — HEAT.   193

Q. 5. Calling the pressure of one atmosphere 1000^g per square centimeter, we have the proportion
$$900 : 1000 :: 1000 : 1111\tfrac{1}{9}^{cc}.$$

Q. 7. $1000 \times 1 : 1 \times 200 :: 273 : 54.6°$ absolute temperature, or $-218.4°$ C.

Q. 8.

Melting Points.		Boiling Points.	
Alcohol	Always liquid	Carbonic acid	195° C.
Mercury	234.2° C.	Ammonia	233°
Sulphuric acid	238.6°	Sulphurous acid	263°
Ice	273°	Ether	308°
Phosphorus	317°	Carbon bisulphide	321°
Sulphur	388°	Alcohol	351°
Tin	about 506°	Water	373°
Lead	607°	Mercury	623°
Zinc	698°		
Silver	1273°		
Gold	1473°		
Cast-iron	1323–1523°		
Wrought-iron	1773–1873°		
Iridium	2223°		

Q. 9. When the barometer is at 30 in., the pressure is 15 lbs. per square inch; and when it is at 29 in., the pressure is $\tfrac{29}{30}$ of 15 lbs. $= 14.5$ lbs. per square inch.

$32°$ F. $= 492°$ F. absolute temperature; $68°$ F. $= 528°$ F. absolute temperature; hence, we may write the compound proportion

$$\left.\begin{array}{r}528 : 492 \\ 15 : 14.5\end{array}\right| :: \frac{1}{25} : \frac{1}{27.7} \text{ lbs.}$$

PAGE 173. Q. 1. To convert 1^k of ice at $0°$ C. into water at $0°$ C. requires 80 calories; to raise it to the boiling point requires 100 calories more, and to change this water into steam, 537 calories are necessary. The total heat that disappears in the change is, therefore, $80 + 100 + 537 = 717$ calories for each kilo, *i.e.*, 71,700 calories for 100^k.

Q. 2. (*a*) In condensing 1000^k of steam at $100°$ C. into liquid at $100°$ C., $1000 \times 537 = 537,000$ calories are liberated;

in falling from 100° C. to 80° C., the heat given off is evidently $1000 \times 20 = 20{,}000$ calories, making a total of 557,000 calories given out to the building.

(b) 1^k of water requires 100 calories to raise it from 0° C. to 100° C.; with 557,000 calories, then, we can raise 5570^k to the same temperature.

Q. 3. 50^k of water at 100° C. can impart 5000 calories to the ice; to melt 1^k of ice at 0° C. takes 80 calories; the amount of ice that may be melted is, then, $\frac{5000^k}{80} = 62.5^k = 137.73 +$ lbs.

Q. 4. $0.504 \times 10 + 80 + 10 = 95.04$ calories.

Q. 5. (a) When the water is at the boiling point, 100° C., 100 calories have been used for each kilo; the ice has been converted into water at 20° C., (b) each kilo having consumed 80 calories in melting.

Q. 7. If we call the resulting temperature $T°$, following the experiment with the sheet lead, we have the equation:—

$$\frac{T}{100 - T} = \text{specific heat of iron} = 0.1138;$$

or $1.1138\, T = 11.38$; whence $T = 10.21° +$ C.

Q. 8. $\frac{5}{95} =$ specific heat $= 0.0526 +$.

Q. 9. 50^k of water at 80° could transmit 4000 calories; but the specific heat of mercury is 0.0333; therefore 50^k of mercury at 80° can transmit only 0.0333 of $4000 = 133.2$ calories. Hence $133.2 \div 80 = 1.665^k$.

CHAPTER IV.

ELECTRICITY AND MAGNETISM.

PAGE 204. Q. 1. From the table (p. 203) we see that the ratio of the relative resistances of iron and copper is $\frac{6.46}{1.06} = 6.09 +$; ∴ $6.09 +$ miles of copper wire offers the same resistance as one mile of iron wire of the same size.

CHAPTER IV.—ELECTRICITY AND MAGNETISM.

Q. 3. By referring again to the table, it is seen that by the addition of the acid the conductivity is made about 5000 times greater.

Q. 4. $\dfrac{2000000}{1.06} = 1,886,792.45+$ times.

Q. 5. Employ plates of large surface, and place them near together.

Q. 6. $R = 9.72 \times \dfrac{500}{14^2} = 24.7+$ ohms.

Q. 7. $1 = 9.72 \times \dfrac{l}{6^2}$; $\therefore l = 3.7+$ ft.

Q. 8. $R = 127.3 \times \dfrac{48}{14^2} = 31.1+$ ohms.

Q. 9. $R = 59.1 \times \dfrac{5280}{175^2} = 10.1+$ ohms.

PAGE 207.

Q. 1. $C = \dfrac{E}{R+r} = \dfrac{1}{10+2} = \tfrac{1}{12}$ ampère.

Q. 2. $C = \dfrac{E}{R+r} = \dfrac{1}{3+3} = \tfrac{1}{6}$ ampère.

Q. 3. $C = \dfrac{E}{R+r} = \dfrac{1.95}{3+3} = 0.325$ ampère.

PAGE 209.

Q. 1. $C = \dfrac{E}{R+r} = \dfrac{10}{200+(0.5 \times 10)} = 0.048+$ ampère.

PAGE 210.

Q. 2. $C = \dfrac{E}{R+r} = \dfrac{40 \times 1.95}{(1500+100)+(0.5 \times 40)}$
$= 0.048+$ ampère.

Q. 3. $R = k\dfrac{l}{d^2} = 59.1 \times \dfrac{5280}{165^2} = 11.4+$ ohms:

$C = \dfrac{E}{R}$; or, $0.02 = \dfrac{E}{0.5+11.4}$; whence

$E = 0.238+$ volts;

therefore less than one cell would be sufficient.

196   SOLUTIONS TO PROBLEMS.

Q. 4.   Arranged in multiple arc,

$$C = \frac{1}{\frac{3}{210}+10} = \frac{210}{2103} = 0.099 + \text{ ampère};$$

arranged in series,

$$C = \frac{210}{(3 \times 210)+10} = 0.32 + \text{ ampères},$$

showing that the latter arrangement is preferable.

Q. 5. In this case the current $C = \dfrac{10E}{10r+R}$; with one cell the current $C = \dfrac{E}{r+R}$. It is clear, then, that if R is very small, the current of one cell is about as strong as that of ten, and the consumption in the other nine is waste.

Q. 6. $C = \dfrac{20}{\dfrac{20 \times 3}{2}+120} = 0.133 + \text{ ampère}.$

Q. 7. We may have any arrangement in which the number of series multiplied by the number of cells in each series equals 30; *e.g.*, 3 rows of 10 cells each; 5 of 6 each, etc. Suppose, first, that the whole 30 are connected in series; then,

(a) $C = \dfrac{30}{(0.8 \times 30)+10} = 0.882 + \text{ ampère};$

if the cells are joined in 2 series of 15 cells each,

(b) $C = \dfrac{15}{\dfrac{0.8 \times 15}{2}+10} = 0.937 + \text{ ampère};$

if in 3 series of 10 cells each,

(c) $C = \dfrac{10}{\dfrac{0.8 \times 10}{3}+10} = 0.789 + \text{ ampère}.$

It is easily seen that with $R = 10$ ohms, the arrangement (b) is best.

Let us take $R = 30$ ohms:

(a) $\quad C = \dfrac{30}{(0.8 \times 30) + 30} = 0.555+$ ampère;

(b) $\quad C = \dfrac{15}{\dfrac{0.8 \times 15}{2} + 30} = 0.416+$ ampère;

(c) $\quad C = \dfrac{10}{\dfrac{0.8 \times 10}{3} + 30} = 0.306+$ ampère.

In this case the arrangement (a) gives the largest value of C.

In general, the best manner of grouping a given number of cells, in order to give the strongest possible current through a given external conductor, is that by which the internal and external resistances are as nearly equal as possible.

Applying this to the example above, we see that the internal resistances (r) in the three arrangements are 24 ohms, 6 ohms, and $2\frac{2}{3}$ ohms; and we found that $r = 6$ gave the best result when $R = 10$; and that $r = 24$ gives the best result when $R = 30$, both of which agree with the principle given above.

CHAPTER V.

SOUND.

PAGE 285. Q. 3. Since the velocity of sound in gases is inversely proportional to the square root of their densities, the densities will be inversely proportional to the squares of the velocities; *i.e.*,

density of carbonic acid : density of air : $331^2 :: 262^2$;

or, the density of carbonic acid is $1.596+$ times that of air.

198   SOLUTIONS TO PROBLEMS.

PAGE 293. Q. 1. A fork vibrating 256 times per second produces wave-lengths twice as long as those produced by a fork vibrating 512 times per second.

Q. 2. Wave-length = 4 times the length of the resonance tube, or $4 \times 22.26^{cm} = 89.04^{cm} = 0.8904^m$. The velocity at 16° C. is 342^m per second; therefore $0.8904 = \dfrac{342}{\text{number of vibrations}}$, whence the required number is $\dfrac{342}{0.8904} = 384.09+$ vibrations.

Q. 3. Wave-length $= \dfrac{342}{384} = 0.89^m+ = 89^{cm}$.

PAGE 301. Q. 1. The vibration number of C'' is 528;
that of $D'' = \tfrac{9}{8}$ of $528 = 594$
" " $E'' = \tfrac{5}{4}$ " " $= 660$
" " $F'' = \tfrac{4}{3}$ " " $= 704$
" " $G'' = \tfrac{3}{2}$ " " $= 792$
" " $A'' = \tfrac{5}{3}$ " " $= 880$
" " $B'' = \tfrac{15}{8}$ " " $= 990$
" " $C''' = 2$ times $528 = 1056$

Q. 2. The vibration number of C_{-1} is $\tfrac{1}{2}$ that of $C = \tfrac{1}{2}$ of $132 = 66$.

Q. 3. Wave-length $= \dfrac{342}{\text{number of vibrations}}$;
∴ wave-length of $C' = \tfrac{342}{264} = 1.29^m+$
" " " $D' = \tfrac{342}{297} = 1.15^m+$
" " " $E' = \tfrac{342}{330} = 1.03^m+$
" " " $F' = \tfrac{342}{352} = 0.97^m+$
" " " $G' = \tfrac{342}{396} = 0.86^m+$
" " " $A' = \tfrac{342}{440} = 0.77^m+$
" " " $B' = \tfrac{342}{495} = 0.69^m+$
" " " $C'' = \tfrac{342}{528} = 0.64^m+$

Q. 4. The whole wave-length of C' we have found to be 129^{cm}; therefore, the length of the resonance tube is $\tfrac{1}{4}$ of $129^{cm} = 32.25^{cm}$.

CHAPTER V. — SOUND.

PAGE 304. Q. 3. It is evident, from the table of Fig. 216, page 300, that the F string is $\frac{3}{4}$ as long as the C string.

PAGE 308. Q. 1. If C and G are sounded simultaneously, there will result for

C, 1, 2, 3, 4, 5, 6, 7, 8, etc. $\Big\}$ times the number of vibra-
G, $\frac{3}{2}$, 3, $\frac{9}{2}$, 6, 9, etc.

tions made by the fundamental of C. From this arrangement, we see that the second overtone of C harmonizes with the first of G; the fifth of C with the third of G, etc.

Q. 2. The notes with their vibration ratios are as follows: —

C	D	E	F	G	A	B	C'
1	$\frac{9}{8}$	$\frac{5}{4}$	$\frac{4}{3}$	$\frac{3}{2}$	$\frac{5}{3}$	$\frac{15}{8}$	2

Coupling C with each of the following, the ratios of their respective vibration numbers are as follows: —

$8:9,\ 4:5,\ 3:4,\ 2:3,\ 3:5,\ 8:15,\ 1:2;$

arranging these on the required principle, we have first, $C\ (1:1)$; then

$C'\ (1:2);\ G\ (2:3);\ F\ (3:4);\ A\ (3:5);$
$E\ (4:5);\ D\ (8:9);\ B\ (8:15).$

CHAPTER VI.

LIGHT.

PAGE 335. Q. 1. A wall of each of the rooms would receive the same quantity of light; viz., $\frac{1}{6}$ of the total amount.

Q. 2. A wall of the first room contains 100 sq. ft., one of the third, 900 sq. ft.; since each wall receives the same total quantity of light, it is clear that the larger one receives only $\frac{1}{9}$ as much as the small one per square foot.

Q. 3. We may look upon the shadow as a pyramid of which the light is the apex and the boards right sections; now, the area of such a section varies as the square of its distance from the apex; the ratio of the distances is 1 : 3; the ratio of the areas is, therefore, 1 : 9.

If the board is withdrawn, the light, intercepted before, will illuminate 900^{qcm} of the screen.

Q. 4. The reason for the law of inverse squares is involved in the answer to the first part of Q. 3 above.

Q. 6. The ratio of the distances is 1 : 4; by the application of the laws of Inverse Squares, the ratio of the intensities at equal distances will be 1 : 16, *i.e.*, the gas flame may be said to have 16 candle-power.

PAGE 354. Q. 2. (*a*) The required relative index equals

$$\frac{\text{absolute index of diamond}}{\text{absolute index of water}} = \frac{2.5}{1.33} = 1.87+.$$

(*b*) The relative index equals $\frac{1.000294}{1.33} = 0.75+.$

www.ingramcontent.com/pod-product-compliance
Lightning Source LLC
Chambersburg PA
CBHW020240090426
42735CB00010B/1780